Brownian Motion
and Potential Theory,
Modern and Classical

Contemporary Mathematics and Its Applications: Monographs, Expositions and Lecture Notes

Print ISSN: 2591-7668
Online ISSN: 2591-7676

Series Editor
M Zuhair Nashed *(University of Central Florida)*

Editorial Board
Guillaume Bal *(University of Chicago)*
Gang Bao *(Zhejiang University)*
Liliana Borcea *(University of Michigan)*
Raymond Chan *(The Chinese University of Hong Kong)*
Adrian Constantin *(University of Vienna)*
Willi Freeden *(University of Kaiserslautern)*
Charles W Groetsch *(The Citadel)*
Mourad Ismail *(University of Central Florida)*
Palle Jorgensen *(University of Iowa)*
Marius Mitrea *(University of Missouri Columbia)*
Otmar Scherzer *(University of Vienna)*
Frederik J Simons *(Princeton University)*
Edriss S Titi *(Texas A&M University)*
Luminita Vese *(University of California, Los Angeles)*
Hong-Kun Xu *(Hangzhou Dianzi University)*
Masahiro Yamamoto *(University of Tokyo)*

This series aims to inspire new curriculum and integrate current research into texts. Its aims and main scope are to publish:

– Cutting-edge Research Monographs
– Mathematical Plums
– Innovative Textbooks for capstone (special topics) undergraduate and graduate level courses
– Surveys on recent emergence of new topics in pure and applied mathematics
– Advanced undergraduate and graduate level textbooks that may initiate new directions and new courses within mathematics and applied mathematics curriculum
– Books emerging from important conferences and special occasions
– Lecture Notes on advanced topics

Monographs and textbooks on topics of interdisciplinary or cross-disciplinary interest are particularly suitable for the series.

Published

Vol. 9   *Brownian Motion and Potential Theory, Modern and Classical*
         by Palle E T Jorgensen, Murali Rao & James Tian

Vol. 8   *Mathematics of Multilevel Systems: Data, Scaling, Images, Signals, and Fractals*
         by Palle E T Jorgensen & Myung-Sin Song

Vol. 7   *Operator Theory and Analysis of Infinite Networks: Theory and Applications*
         by Palle E T Jorgensen & Erin P J Pearse

More information on this series can also be found at
https://www.worldscientific.com/series/cmameln

Contemporary Mathematics and Its Applications
Monographs, Expositions and Lecture Notes

Vol. **9**

Brownian Motion and Potential Theory, Modern and Classical

Palle Jorgensen
University of Iowa, USA

Murali Rao
University of Florida, USA

James Tian
The American Mathematical Society, USA

World Scientific

NEW JERSEY • LONDON • SINGAPORE • BEIJING • SHANGHAI • HONG KONG • TAIPEI • CHENNAI • TOKYO

Published by

World Scientific Publishing Co. Pte. Ltd.
5 Toh Tuck Link, Singapore 596224
USA office: 27 Warren Street, Suite 401-402, Hackensack, NJ 07601
UK office: 57 Shelton Street, Covent Garden, London WC2H 9HE

Library of Congress Control Number: 2024043041

British Library Cataloguing-in-Publication Data
A catalogue record for this book is available from the British Library.

Contemporary Mathematics and Its Applications:
Monographs, Expositions and Lecture Notes — Vol. 9
BROWNIAN MOTION AND POTENTIAL THEORY, MODERN AND CLASSICAL

Copyright © 2025 by World Scientific Publishing Co. Pte. Ltd.
All rights reserved. This book, or parts thereof, may not be reproduced in any form or by any means, electronic or mechanical, including photocopying, recording or any information storage and retrieval system now known or to be invented, without written permission from the publisher.

For photocopying of material in this volume, please pay a copying fee through the Copyright Clearance Center, Inc., 222 Rosewood Drive, Danvers, MA 01923, USA. In this case permission to photocopy is not required from the publisher.

ISBN 978-981-12-9431-0 (hardcover)
ISBN 978-981-12-9477-8 (paperback)
ISBN 978-981-12-9432-7 (ebook for institutions)
ISBN 978-981-12-9433-4 (ebook for individuals)

For any available supplementary material, please visit
https://www.worldscientific.com/worldscibooks/10.1142/13872#t=suppl

Desk Editors: Nambirajan Karuppiah/Lai Fun Kwong

Typeset by Stallion Press
Email: enquiries@stallionpress.com

*Dedicated to
the memory of Professor Kiyosi Itô.*

Preface to the Second Edition

While Brownian motion and potential theory might seem like two seemingly unrelated branches of mathematics/physics, it has emerged over time that there are close and exciting mathematical interconnections. Indeed, by now these interconnections have had profound implications — they have helped foster independent advances, and they have demonstrated how tools from diverse areas apply to the others. In recent years, these interconnections have further led to new outlooks and to new insight. The resulting subject is now often referred to as probabilistic potential theory. The focus of this book is twofold: one is foundations and the other is new directions.

Some of the more traditional applications of ideas from potential theory include a variety of boundary value problems. Historically, the first instances of potentials were perhaps motivated by the study of elliptic partial differential operators, but by now the initial ideas have moved to a host of new and diverse areas of both pure and applied mathematics, e.g., complex domains in one or several complex variables, dynamical systems realized on a variety of choices of path space, commutative and non-commutative manifold analysis, to mention just a few. Looking back, we note that an early focus for both classical and abstract potential theory was the study of Newton and logarithmic potentials for partial differential equations of elliptic type: e.g., the Laplace equation in regions, free of the masses, generating the potentials, and the Poisson equation in regions occupied by the masses. Thus, in these early cases, the boundary value problems of potential theory arose primarily for elliptic equations and systems. Other early influences include the use of balayage methods (French for sweeping to the boundary), dating back to H. Poincaré. In a rough sketch, for domains D, this amounts to finding a "swept-out" mass distribution on some choice of boundary Γ. Then the potential for this will coincide with the potential of given masses in the interior of D. Yet other applications include Bessel potentials, nonlinear potentials, and Riesz potentials.

This book aims to reach multiple audiences, ranging from students to specialties, and it makes connections to both pure and applied math, and to neighboring areas. The subject of potential theory and its probabilistic counterparts is part of classical analysis and its neighboring fields, and it has now recently gained much wider attention and interest. Indeed, in recent years, many new connections have emerged to stochastic calculus, stochastic processes, dynamical systems, PDEs, analysis of boundaries, the theory of kernels, and also a variety of applications, including in physics, actuarial science, and financial mathematics. The present focus is on the core foundational themes of the subject. Our book grew out from an initial informal set of lecture notes for a graduate course, by the second named author, Prof Murali Rao. Here we aim to systematically cover both the foundations, as well as to point readers toward many new trends.

A word about the first edition, since it appeared in 1977 in the Institute of Mathematics Research Publications Department's publication series archive as Lecture Notes Series Number 47, Aarhus University Ny Munkegade 118, DK-8000 Aarhus, it has only existed in an informal pre-tex-typed version. And the number of existing paper versions was small. Nonetheless, it was cited around the world, and it is linked in such websites as https://www1.essex.ac.uk/maths/people/fremlin/rao.htm (University of Essex).

Two years ago, Palle Jorgensen contacted Murali Rao, suggesting a second edition for publication. He is extremely grateful for Prof Murali Rao's warm reception. He is also grateful to Prof Jacob Schach Møller, Head of Department of the Aarhus University Mathematics Institute, both for giving permission for use of the original and for his enthusiastic encouragement. We should add that, over the intervening long period, the subject has in fact grown in its significance, both as a core area of pure and applied mathematics and in branching out to probability, statistics, stochastic analysis, financial mathematics, and physics and adding to the role it plays in the study of boundaries for theory diffusion processes. The original 1977 version by Prof Murali Rao was reviewed in MathSciNet: Rao, Murali Brownian motion and classical potential theory. Lecture Notes Series, No. 47, Aarhus Universitet, Matematisk Institut, Aarhus, 1977. Quoting from the review by F. B. Knight: "This set of 'lecture notes' may be described as a collection of 140 well-polished exercises, and a like number of examples, attached to a relatively sparse text. The author has done a commendable job of assembling many facts and theorems, and of presenting them with rigor and elegance." The review goes on with a chapter-by-chapter summary, concluding, "The final chapter, Potential theory, covers the entire range from Newtonian to abstract axiomatic, and including 'a little something on Dirichlet spaces' in 70 pages."

The main additions in this 2nd edition fall into three categories: (i) some updates inside the 1st edition text, (ii) many updates to the cited literature, both research papers and books, and finally, (iii) a new appendix where a more general framework is offered.

Palle Jorgensen
Department of Mathematics, University of Iowa, USA
palle-jorgensen@uiowa.edu

Murali Rao
Department of Mathematics, University of Florida, Gainesville, USA
mrao@ufle.edu

James Tian
Mathematical Reviews, 416 4th Street Ann Arbor, MI, USA
jft@ams.org

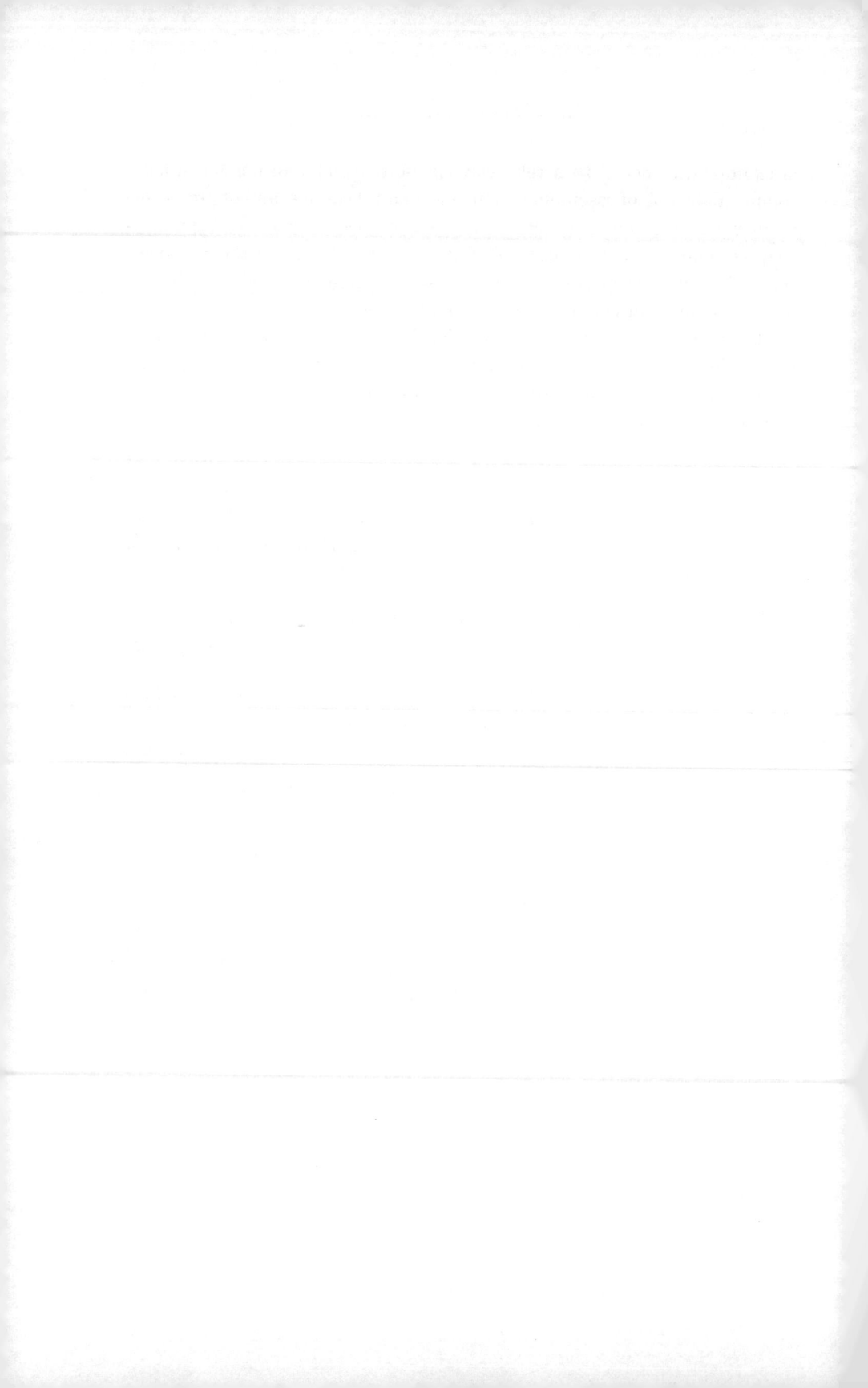

Preface to the First Edition

In these notes, we present some basic classical potential theory. Although the language of Brownian motion is used, the discussion is more analytic than probabilistic.

Chapters 5–8 describe the principal results, whereas the first three chapters are of a more or less general character. Ample hints are provided for almost all the exercises.

Senior undergraduate-level mathematics is a necessary prerequisite for reading this set of notes.

Murali Rao

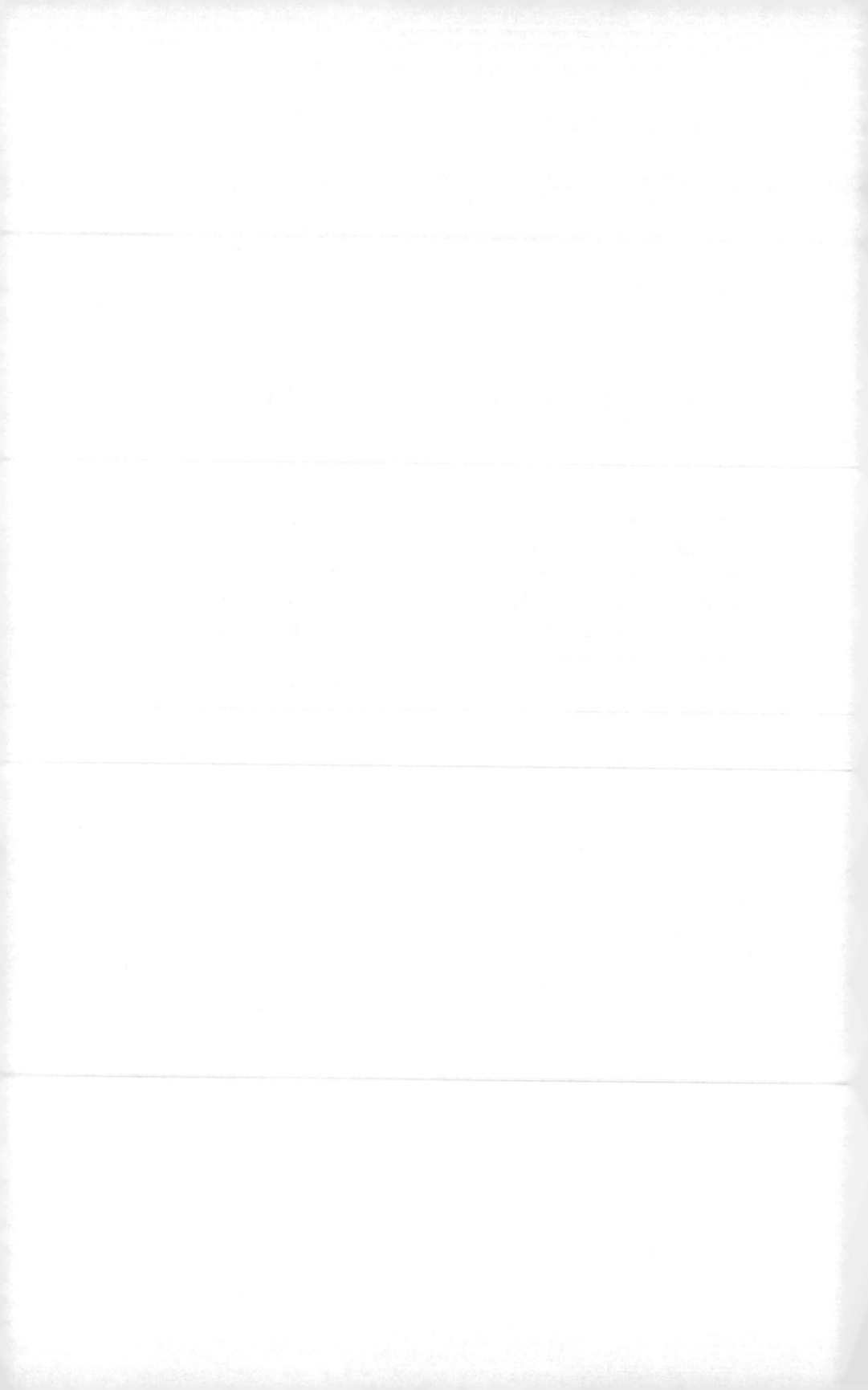

About the Authors

Palle Jorgensen is a Professor of Mathematics at the University of Iowa, USA, and a Fellow of the American Mathematical Society. He has a distinguished teaching career, having previously taught at Stanford University, Aarhus University, and the University of Pennsylvania. Jorgensen has directed 35 PhD theses and is a highly cited research mathematician with numerous awards, including a Faculty Fellow award at the University of Iowa. In 2018, he was selected as the NSF/CBMS speaker, delivering a series of lectures published in the AMS/CBMS book series. He has authored over 300 research papers and 15 books, covering both pure and applied mathematics, with research interests in operator algebras, harmonic analysis, signal processing, wavelets, and mathematical physics.

Murali Rao is a Professor at the Mathematics Department of the University of Florida, Gainesville, with a PhD supervised by Kiyosi Ito. His teaching and research span mathematics, probability, and stochastic processes. Rao's research is interdisciplinary, focusing on areas such as probability theory, density estimation, white noise analysis, and various mathematical tools and concepts related to stochastic processes.

James Tian is an Associate Editor at *Mathematical Reviews* in Ann Arbor, Michigan.

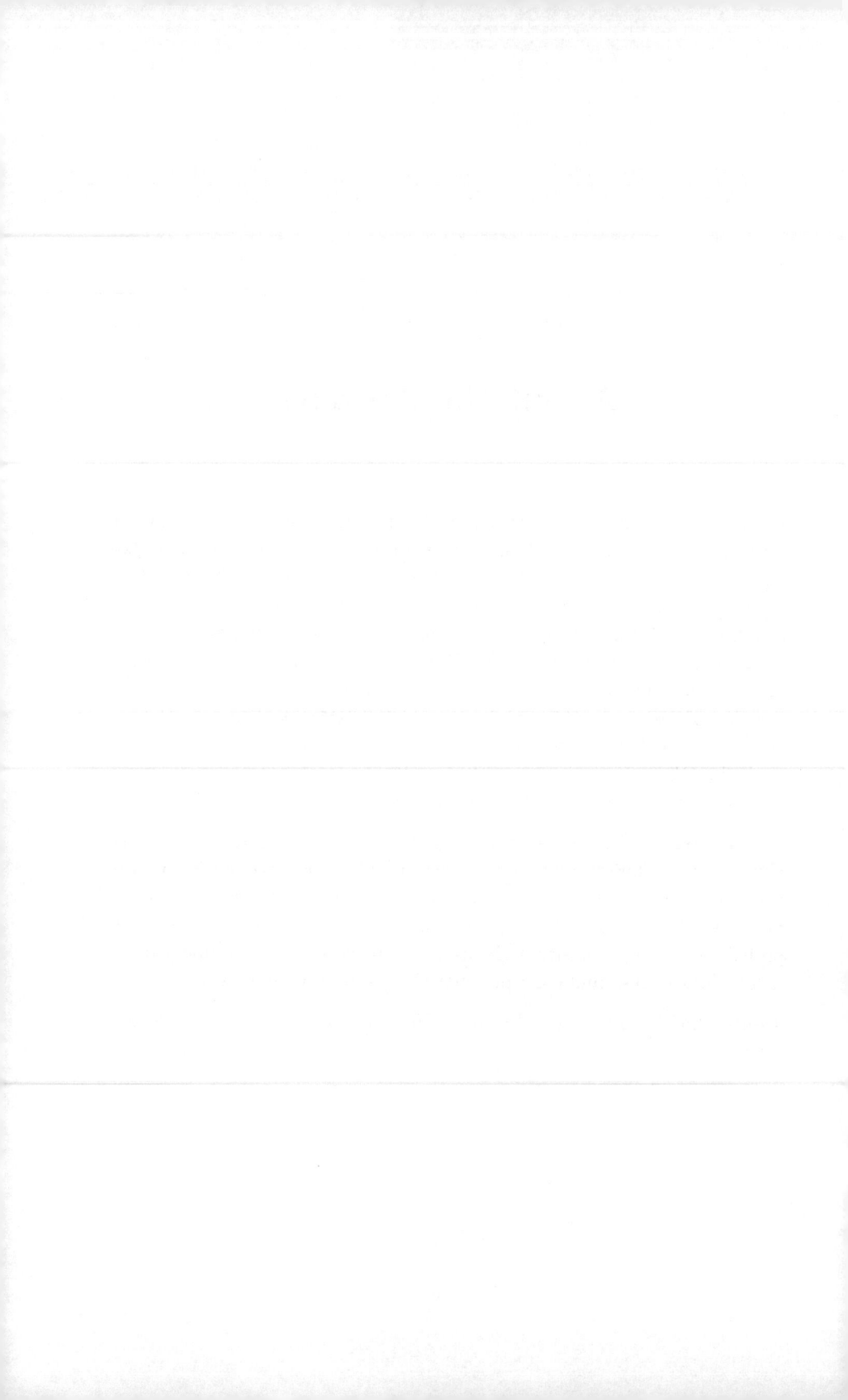

Contents

Preface to the Second Edition   vii
Preface to the First Edition   xi
About the Authors   xiii

1 Introduction   **1**

2 Martingales and Markov Processes   **5**
    2.1 Optional Sampling, Inequalities, and Convergence   5
    2.2 Continuous Parameter   13

3 Brownian Motion and Ito Calculus   **19**
    3.1 The d-Dimensional Brownian Motion   20
    3.2 Strong Markov Property   24

4 Semi-Groups of Operators, Potentials,
    and Diffusion Equations   **43**
    4.1 Semi-Groups   45
    4.2 Infinitesimal Generators   50
    4.3 Potential Operators   60

5 Harmonic Functions, Dynkin, and Transforms   **71**
    5.1 Dynkin's Formula   72
    5.2 Dirichlet Problem   78
    5.3 The Kelvin Transformation   88
    5.4 Boundary Limit Theorems of Fatou   93
    5.5 Spherical Harmonics   98

6 Superharmonic Functions and Riesz Measures — 109
- 6.1 Superharmonic Functions — 110
- 6.2 Applications — 119
- 6.3 Riesz Measure — 123
- 6.4 The Continuity Principle — 133
- 6.5 The Dirichlet Problem Revisited — 137

7 Green Functions, Boundary Value Problems, and Kernels — 143
- 7.1 Green Functions for Bounded Open Sets — 144
- 7.2 Unbounded Open Subsets of $\mathbb{R}^2$ — 158
- 7.3 Unbounded Open Sets *(Continued)* — 164
- 7.4 Examples — 169
- 7.5 The Green Function and Relative Transition — 176

8 Potential Theory, Capacity, Boundaries, Dirichlet Spaces, and Applications — 183
- 8.1 Some Potential Theoretic Principles — 184
- 8.2 Capacity — 198
- 8.3 Applications — 205
- 8.4 Balayage — 209
- 8.5 Dirichlet Spaces — 214

Appendix Kernels and More General Classes of Gaussian Processes — 233
- A.1 Positive Definite Kernels — 233
- A.2 Gaussian Processes — 236
- A.3 Sigma-Finite Measure Spaces and Gaussian Processes — 238

Index — 243

Chapter 1

Introduction

The present outlook for potential theory as covered in this book is inclusive, ranging from classical to modern, and from analytic to probabilistic, thus covering the range from Newtonian to abstract/axiomatic potential theory, including Dirichlet spaces. A leading theme is the path taking the reader from stochastic analysis with Brownian motion at the beginning of this book, continuing through this book, and reaching potential theory in Chapter 8. This path thus takes the reader through the following themes: martingales, diffusion processes, semi-groups and potential operators, analysis of superharmonic functions, Dirichlet problems, balayage, boundaries, and Green functions. All themes form the focus of the intervening chapters in this book.

While this chapter's dependences lead naturally from Chapters 2 to 8, some readers may find it natural to start reading first any one of the seven chapters. The interdependences are sketched in the following dependency diagram (Figure 1.1).

This volume is motivated by several themes: (i) core questions in classical and modern potential theory, (ii) pointing to new directions, and (iii) putting in perspective the tools from probability (of Markov transitions) and stochastic analysis. Indeed, the tools from stochastic analysis have proved to be of independent interest in both pure and applied mathematics, including harmonic analysis, dynamical systems theory, PDE, and financial mathematics, e.g., hedging and pricing formulas for derivative securities.

A key tool in our results in Chapter 8 is balayage, i.e., the mathematics of "sweeping to the boundary". While the corresponding balayage operators originally were designed for applications to problems in potential theory, they have subsequently found uses in many other new and surprising

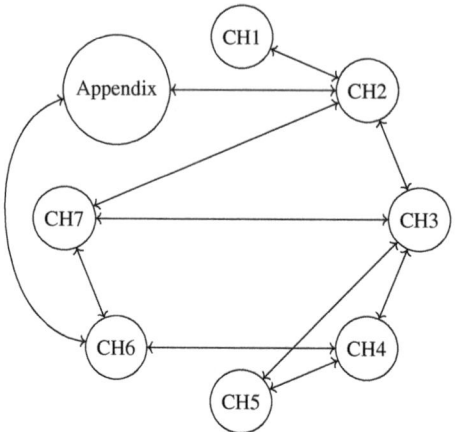

Fig. 1.1 Interdependences of chapters.

areas of mathematics, for example in obtaining geometric information about quadrature domains and in the derivation of solutions to certain moving boundary problems (e.g., for Hele–Shaw flows). Other applications include dynamics, the determination of equilibrium measures, specific Martin boundaries, harmonic analysis, analytic continuation, the analysis of quadrature domains, isoperimetric inequalities, infinite networks, the study of restriction operators, and doubly orthogonal systems of analytic functions. For these applications, we refer to [1–9].

References

[1] J. J. Benedetto and M. J. Begué. Fourier operators in applied harmonic analysis. In *Sampling Theory, A Tenaissance*. Applied and Numerical Harmonic Analysis, pp. 185–215. Birkhäuser/Springer, Cham, 2015.

[2] B. Gustafsson, M. Putinar, E. B. Saff and N. Stylianopoulos. Bergman polynomials on an archipelago: Estimates, zeros and shape reconstruction. *Adv. Math.*, 222(4): 1405–1460, 2009.

[3] B. Gustafsson, M. Putinar and H. S. Shapiro. Restriction operators, balayage and doubly orthogonal systems of analytic functions. *J. Funct. Anal.*, 199(2): 332–378, 2003.

[4] B. Gustafsson and M. Sakai. Properties of some balayage operators, with applications to quadrature domains and moving boundary problems. *Nonlinear Anal.*, 22(10): 1221–1245, 1994.

[5] B. Gustafsson and M. Sakai. On potential-theoretic skeletons of polyhedra. *Geom. Dedicata*, 76(1): 1–30, 1999.

[6] B. Gustafsson, M. Sakai and H. S. Shapiro. On domains in which harmonic functions satisfy generalized mean value properties. *Potential Anal.*, 7(1): 467–484, 1997.
[7] F.-Y. Maeda. Duality of two balayage-space structures and Dirichlet integrals on balayage spaces. *Potential Anal.*, 4(6): 595–613, 1995.
[8] M. Sakai. Restriction, localization and microlocalization. In *Quadrature Domains and Their Applications*. Operator Theory: Advances and Applications, Vol. 156, pp. 195–205. Birkhäuser, Basel, 2005.
[9] N. Zorii. Harmonic measure, equilibrium measure, and thinness at infinity in the theory of Riesz potentials. *Potential Anal.*, 57(3): 447–472, 2022.

Chapter 2

Martingales and Markov Processes

In broad outline, a martingale X_t refers to a stochastic process indexed by time. In addition, a time filtration F_t is introduced for the underlying probability space. The martingale condition for the process X_t refers to present, say t, and future, T. It states that X_T conditioned on F_t is the value X_t. Martingale was introduced by Paul Lévy in 1934 and by Ville in 1939. Much of the subsequent theory was done by Joseph Leo Doob who also introduced the notion of successful betting strategies in games of chance.

The original material in Chapter 1 dealing with martingales is still current, and we have added a discussion of new directions. In addition, for the benefit of readers, we offer the following citations covering new directions [1–3, 6–13, 15–19].

The new directions include relative entropy between martingale diffusions on the line, prediction for positive self-similar Markov processes, fractional Brownian motion, finite variation processes, maximum for sums of random variables, and risk theory, and second-order processes.

For the benefit of beginner readers, we offer the following supplementary introductory text [4].

2.1 Optional Sampling, Inequalities, and Convergence

We will not be needing too much martingale theory. What little is done is to make the notes more complete. In the following, we assume given a fixed probability space $(\Omega, \mathscr{B}, \mathbb{P})$. All σ-fields considered are *sub-σ-fields of $\mathscr{B}$*. In some of the examples, some knowledge of the relevant concepts is needed.

Let $F_0, F_1, \ldots$ be a (finite of infinite) sequence of σ-fields and $X_0, X_1, \ldots$ a sequence of random variables. $\{X_i\}$ is said to be *adapted* to

F_i if X_i is F_i-measurable. An integer-valued (possibly ∞) random variable T is called a *stopping time* relative to F_i if $(T = i) \in F_i$ for all i. For a stopping time T, the σ-field F_T consists of the events A for which $A \cap (T = i) \in F_i$ for all i. If $F_0 \subset F_1 \subset \cdots$ is increasing and $T \geq S$ are stopping times, then $F_T \supset F_S$; if F_i are decreasing and $T \geq S$ are stopping times, then $F_T \subset F_S$. These are easily verified. *Until further notice, F_i will be an increasing sequence of σ-fields.*

A sequence $\{X_i\}$ of random variables having expectations and adapted to $\{F_i\}$ is called a *super martingale* if

$$\mathbb{E}\left[X_{i+1} \mid F_i\right] \leq X_i, \quad i = 0, 1, 2, \ldots, \tag{2.1.1}$$

a *sub-martingale* if $\{-X_i\}$ is a super martingale, and a *martingale* if both $\{X_i\}$ and $\{-X_i\}$ are super martingales. Thus, for a sub-martingale, the inequality in (2.1.1) is reversed and for a martingale, the inequality in (2.1.1) is actually an equality. The expression "let $\{X_i\}$ be a super martingale" will mean that $\{X_i\}$ is adapted to $\{F_i\}$, X_i have expectations, and $\{X_i\}$ is a super martingale. A super martingale is called *non-negative if all X_i are non-negative*.

Denoting by d_i the difference $X_i - X_{i-1}$ and defining $X_{-1} = 0$, (2.1.1) says that the conditional expectation of d_i given F_i is less or equal to zero. Therefore, if $a_0 = b_0$ and for $i \geq 1$, $a_i \leq b_i$ are bounded and F_{i-1} measurable, then

$$\mathbb{E}\left[a_i d_i\right] \geq \mathbb{E}\left[b_i d_i\right]. \tag{2.1.2}$$

We get the following:

Proposition 2.1.1. *Let $\{X_i, i \geq 0\}$ be a super martingale. If $a_0 = b_0$ and for $i \geq 1$, $a_i \leq b_i$ are bounded F_{i-1}-measurable functions, then for any N,*

$$\mathbb{E}\left[\sum_0^N b_i d_i\right] \leq \mathbb{E}\left[\sum_0^N a_i d_i\right], \tag{2.1.3}$$

where $d_i = X_i - X_{i-1}$ and $X_{-1} = 0$.

Suppose now $T \leq S$ are stopping times. Taking $a_i = $ indicator of $(T \geq i) \leq b_i = $ indicator of $(S \geq i)$, we get from (2.1.3) the following:

Corollary 2.1.2 (Optional Sampling Theorem). *Let $\{X_i, i \geq 0\}$ be a supermartingale. If $T \leq S$ are stopping times, then for any N,*

$$\mathbb{E}\left[X_T : T \leq N\right] \geq \mathbb{E}\left[X_S : S \leq N\right] + \mathbb{E}\left[X_N : T \leq N < S\right]. \tag{2.1.4}$$

In case of a martingale, there is equality in (2.1.4).

Remark 2.1.3. There is a simple way of getting a more general conditional inequality from (2.1.4). For $A \in F_T$, let T_A be the stopping time which is T on A and infinite otherwise. Similarly for S_A. An application of (2.1.4) to these stopping times leads to the said conditional form of (2.1.4). This is clearly a general trick and it is useful to make a note of it. For some applications of the optional sampling theorem, see Karlin–Taylor [5, pp. 263–272].

If X_i is non-negative, the last term in (2.1.4) may be omitted. Letting N tend to infinity, we obtain by monotone converge theorem $\mathbb{E}[X_T] \geq \mathbb{E}[X_S]$, where we put $X_\infty = 0$. This already implies that a non-negative super martingale converges almost surely. See Exercises 2.1.5 and 2.1.6.

The case of a decreasing sequence of σ-fields is equally important. Suppose $F_0 \supset F_1 \supset F_2 \supset \cdots$ is a decreasing sequence of σ-fields. A super martingale relative to F_n will be a sequence $X_0, X_1, \ldots$, where X_n is F_n-measurable, has expectation, and $\mathbb{E}[X_n \mid F_{n+1}] \leq X_{n+1}$. Exactly as before, we obtain an analogue of Corollary 2.1.2: If $T \leq S$ are stopping times, then for any N,

$$\mathbb{E}[X_T : T \leq N] \leq \mathbb{E}[X_S : S \leq N] + \mathbb{E}[X_N : T \leq N < S]. \qquad (2.1.5)$$

We now give another famous inequality of Doob, originally proved to establish the martingale convergence theorem. Before we do this, however, we must establish some notation.

Let $\alpha_1, \ldots, \alpha_n$ be an ordered set of numbers and $r < s$ numbers. We say that the set $\{\alpha_1, \ldots, \alpha_n\}$ experiences at least k upcrossings of the interval $[r, s]$ if there exist indices $1 \leq i_1 < j_1 < i_2 < j_2 < \cdots < i_k < j_k \leq n$ such that $\alpha_{i_1} < r$, $\alpha_{j_1} > s$, $\alpha_{i_2} < r$, $\alpha_{j_2} > s$, etc. $\alpha_{i_k} < r$, $\alpha_{j_k} > s$. Graphically, this means the following: suppose we join the successive points (i, α_i) by a broken line, the graph looks like Figure 2.1, and there must be k points above the line s and k points below the line r.

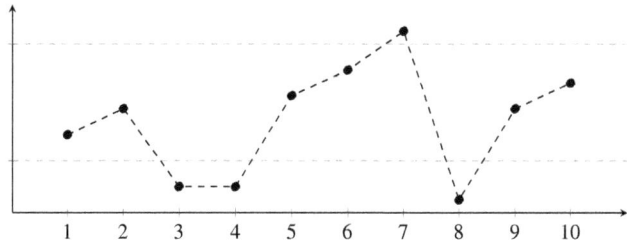

Fig. 2.1 k upcrossings.

Suppose now that $(x_n, n \geq 0)$ is a supermartingale relative to the increasing sequence $(F_n, n \geq 0)$ of σ-fields. If $\gamma < s$ are numbers, let for $i > 1$

$$b_i = \begin{cases} 1 & \text{if there is } j < i \text{ such that } X_j < \gamma \text{ and } X_{j+1}, \ldots, X_{i-1} \leq s, \\ 0 & \text{otherwise.} \end{cases}$$

If $U(N)$ is the number of upcrossings of $[\gamma, s]$ by $X_0, \ldots, X_n$, the reader is invited to convince himself that

$$\sum_1^N b_j d_j \geq (s - \gamma) U(N) + X_n - a \geq (s - \gamma) U(N) - (X_N - a)^-.$$

Taking $a_i \equiv 0 = b_0$ in (2.1.3), we obtain

$$(s - \gamma) \mathbb{E}(U(N)) \leq \mathbb{E}\left[(X_N - \gamma)^-\right]. \tag{2.1.6}$$

Now if $\{F_n, n \leq 0\}$ is decreasing, the σ-fields $G_n = F_{N-n}$ are increasing for $n \leq N$. The following corollary is thus clear from (2.1.6) as N tends to infinity.

Corollary 2.1.4 (Doob's Upcrossing Inequality). *Let X_n be a supermartingale relative to a monotone sequence of σ-fields. If U is the number of upcrossings of $[\gamma, s]$ by the sequence X_n, then*

$$(s - \gamma) \mathbb{E}[U] \leq \sup_N \mathbb{E}[|X_n - \gamma|].$$

In particular, the number of upcrossings is finite with probability 1 provided $\{X_n\}$ is L^1-bounded, i.e., $\sup_n \mathbb{E}[|X_n|] < \infty$.

As a simple corollary, we have the following famous:

Theorem 2.1.5 (Martingale Convergence Theorem). *Let X_n be an L^1-bounded super martingale relative to a monotone sequence of σ-fields. Then $\lim X_n$ exists almost surely.*

Proof. If $Z = \limsup X_n$ and $Y = \liminf X_n$, then $\mathbb{P}[\lim X_n \text{ exists}] = 1$ if and only if $\mathbb{P}[Y < \gamma < s < Z] = 0$ for all rationals $\gamma < s$. Since the event $(Y < \gamma < s < Z)$ implies an infinite number of upcrossings of $[\gamma, s]$ by the sequence $\{X_n\}$, the result is clear from Corollary 2.1.4. $\square$

There are many applications of the martingale convergence theorem. For a nice account of these, we refer to Chapter VIII of Meyer [14].

Example 2.1.6. Consider the probability space $[0,1)$ with the Lebesgue measure. Let F_n = the field generated by the sets $A_{i,n} = [i2^{-n}, (i+1)2^{-n})$, $i = 0, 1, 2, \ldots, 2^n - 1$. $F_n \subset F_{n+1}$ and $\cup F_n$ generates the Borel field in $[0,1)$. Let m be a probability measure on $[0,1)$. Put $X_n = \sum_i 2^n m(A_{i,n}) A_{i,n}$, where $A_{i,n}$ denotes also the indicator function of the set $A_{i,n}$. X_n is a martingale relative to F_n and is L^1-bounded. $Y = \lim X_n$ exists almost everywhere by Theorem 2.1.5.

For any $k > 0$, let $B_k = \{\sup X_n \le k\}$. Let us show that on B_k, $m = Y dx$, where dx is the Lebesgue measure. If $D_i = \{X_j \le k, j \le i\}$, then $m(D_i) = \int_{D_i} X_i dx$. On D_i, $X_i \le k$ so that by dominated convergence, $m(B_k) = \int_{B_k} Y dx$. On the other hand, for any $A \in F_n$ and all $i \ge n$, $m(A) = \int_A X_i dx$ so that by Fatou, $m(A) \ge \int_A Y dx$. Thus, $m = Y dx$ on $(\sup_n X_n < \infty) = \cup B_k$; because X_n converges almost everywhere, this set has probability 1. The measure $m - Y dx$ sits on a set of measure zero, so it is singular. We have obtained Lebesgue's decomposition: $m = Y dx + s$, where s is singular.

The following example assumes some acquaintance with some relevant concepts.

Example 2.1.7. Let D be a bounded domain in $\mathbb{R}^d$ and let us show that a positive harmonic function on D, which is extreme among positive harmonic functions on D, cannot be bounded. Let u be a bounded positive harmonic function on D. Let D_n be open relatively compact subsets of D increasing to D. X_t will denote the Brownian motion in $\mathbb{R}^d$ and T_n, T the exit times from D_n and D, respectively. $u(X_{T_n})$ is then a bounded martingale relative to $\mathbb{P}_a$ and B_{T_n} (the stopped Borel fields) for every $a \in D$. If $F = \limsup u(X_{T_n})$, $\lim u(X_{T_n}) = F$, $\mathbb{P}_a$-almost everywhere for each $a \in D$ by Theorem 2.1.5:

$$u(a) = \mathbb{E}_a[u(X_{T_n})] = \mathbb{E}_a[F].$$

Let $0 \le f \le 1$ be continuous on $\mathbb{R}^d$. The functions $\mathbb{E}.[fu(X_{T_n})]$ are harmonic in D_n and converge. The limit $v = \mathbb{E}.[f(X_{T_n})F]$ is positive harmonic in D. For the same reason, $w = \mathbb{E}.[(1-f)(X_{T_n})F]$ is positive harmonic in D and $u = \frac{1}{2} \cdot 2v + \frac{1}{2} \cdot 2w$. For $0 \le f < \frac{1}{2}$, it is clear that $2v$ cannot be equal to u, i.e., u is not extreme.

If in Example 2.1.6 m is singular relative to Lebesgue measure, Y must vanish almost everywhere. In particular, X_n cannot converge in L^1. Therefore, L^1 convergence is not implied by Theorem 2.1.5.

Definition 2.1.8. A family $\{X_i\}$ of random variables is called *uniformly integrable* if
$$\mathbb{E}\left[|X_i| : |X_i| \geq N\right] \tag{2.1.7}$$
is uniformly small provided N is large enough.

For an equivalent definition, see Exercise 2.1.6.

Proposition 2.1.9. *Let $\{X_n\}$ be a sequence of random variables. Suppose that $\lim X_n = X$ almost everywhere or just in probability. Then X_n tends to X in L^1 if and only if $\{X_n\}$ is uniformly integrable.*

Proof. We will be brief. Suppose $\{X_n\}$ is uniformly integrable. This implies in particular that $\mathbb{E}[|X_n|]$ is uniformly bounded. By Fatou, X is integrable. X_n truncated at N tends to X truncated at N almost everywhere and in L^1 by dominated convergence. The rest is uniformly small by (2.1.7) for large enough N.

The other direction is equally simple. □

In view of the above proposition, we can say L^1-*convergence obtains in Theorem 2.1.5 if and only if $\{X_n\}$ is uniformly integrable.*

It is easy to see that any one of the following implies uniform integrability of a family $\{X_i\}$:

- X_i are bounded by a integrable function.
- The pth moment of X_i is uniformly bounded for some $p > 1$.
- There is a random variable X with finite expectation such that each X_i is the conditional expectation (relative to some σ-field) of X.

Supermartingales relative to increasing and decreasing sequences of σ-fields are not completely similar as the following simple proposition shows. We shall also be using this proposition in the following section.

Proposition 2.1.10. *Let X_n be a supermartingale relative to a decreasing sequence of σ-fields F_n. Then X_n is uniformly integrable if and only if $\sup \mathbb{E}[X_n] < \infty$.*

Proof. $\mathbb{E}[X_n]$ is increasing with n. And if $m \leq n$, $Y_{n,m} = X_m - \mathbb{E}[X_n \mid F_m] \geq 0$. Also $\mathbb{E}[Y_{n,m}] = \mathbb{E}[X_m] - \mathbb{E}[X_n]$ tends to zero as $n \leq m$ tend to infinity. We have, if $A = (|X_m| \leq \lambda)$,
$$\mathbb{E}[|X_m| : A] \leq \mathbb{E}[Y_{n,m}] + \mathbb{E}[|\mathbb{E}(X_n \mid F_m)| : A] \leq \mathbb{E}[Y_{n,m}] + \mathbb{E}[|X_n| : A].$$
$\mathbb{E}[Y_{n,m}]$ is small if n is large enough, uniformly in $m \geq n$. $\mathbb{P}(A)$ will be small for large λ, if we show that $\{X_m\}$ is L^1-bounded, i.e., if $\sup E[|X_m|] < \infty$

and then the last quantity in the above inequality will also be small uniformly in $m \geq n$. Now $\mathbb{E}\left[X_0 : X_m < 0\right] \leq \mathbb{E}\left[X_m : X_m < 0\right]$ so that $\mathbb{E}\left[X_m^-\right]$ is bounded by $\mathbb{E}\left[|X_0|\right]$. Hence, $\sup \mathbb{E}\left[X_m\right] < \infty$ is equivalent to saying that $\{X_m\}$ is L^1-bounded. $\square$

The following "maximal" inequality extends the well-known Kolmogorov inequality.

Lemma 2.1.11. Let X_n, $n = 0, 1, 2, \ldots$, be a non-negative sub-martingale. Then for any $a > 0$

$$a\mathbb{P}\left(\sup_i X_i \geq a\right) \leq \sup_n \mathbb{E}\left[X_n\right]. \tag{2.1.8}$$

Proof. Put $T = \min\left(n : X_n \geq a\right)$, $T = \infty$, if there is no such n. Take $S = \infty$ in Corollary 2.1.2 and remember that we now have a sub-martingale: For any N,

$$\mathbb{E}\left[X_T : T \leq N\right] \leq \mathbb{E}\left[X_N : T \leq N\right].$$

Since $X_T \geq a$, if $T < \infty$, the above inequality implies that

$$a\mathbb{P}\left(T \leq N\right) \leq \mathbb{E}\left[X_N : T \leq N\right]. \tag{2.1.9}$$

(2.1.8) is a consequence of (2.1.9) because the set $(\sup X_i \geq a)$ is the same as the set $(T < \infty)$. $\square$

Exercises

2.1.1. If X_i and Y_i are super martingales so is $X_i + Y_i$. If X_i is a martingale $|X_i|$ is a sub-martingale. If X_i is a non-negative sub-martingale so is X_i^p, $p \geq 1$.

Hint: Hölders inequality.

2.1.2. A supermartingale X_i is called L^1-bounded if $\sup \mathbb{E}\left[|X_i|\right] < \infty$. Show that a martingale is the difference of two non-negative martingales if and only if it is L^1-bounded.

Hint: Suppose X_i is L^1-bounded. For n fixed, $Y_{p,n} = \mathbb{E}\left[|X_{p+n}| \mid F_n\right]$ dominates $|X_n|$ and increases with p. Then $Y_n = \lim Y_{p,n}$ is a martingale.

2.1.3. Show that a supermartingale is the difference of two non-negative super martingales if and only if it is L^1-bounded.

Hint: If X_i is an L^1-bounded supermartingale, for fixed n, $Y_{p,n} = \mathbb{E}\left[X_{p+n} \mid F_n\right]$ is decreasing in p, is dominated by X_n. $Y_n = \lim Y_{p,n}$ is an L^1-bounded martingale. Now use Exercise 2.1.2.

2.1.4. If X_i is a non-negative supermartingale, then $\lim X_i$ exists almost everywhere.

Hint: Given $0 < a < b$, define the sequences T_i, S_i of stopping times by

$$T_1 = \inf\{n : X_n < a\},$$
$$S_1 = \inf\{n : n > T_1, X_n > b\},$$

and inductively

$$T_{n+1} = \inf\{m : m > S_n, X_m < a\},$$
$$S_{n+1} = \inf\{m : m > T_{n+1}, X_m > b\},$$

the infimum over an empty set being always defined ∞. Then $T_n \leq S_n \leq T_{n+1}$, $T_n < \infty$ implies $X_{T_n} < a$ and $S_n < \infty$ implies $X_{S_n} > b$. And

$$A = \cap (S_i < \infty) = \cap (T_i < \infty) = (\liminf X_i < a < b < \limsup X_i).$$

Finally, $b\mathbb{P}[S_i < \infty] \leq \mathbb{E}[X_{S_i}] \leq \mathbb{E}[X_{T_i}] \leq a\mathbb{P}[T_i < \infty]$. Let i tend to ∞ to conclude that $b\mathbb{P}(A) \leq a\mathbb{P}(A)$.

2.1.5. An L^1-bounded supermartingale converges almost surely.

Hint: Use Exercises 2.1.3 and 2.1.4.

2.1.6. Show that uniform integrability of a family $\{X_\alpha\}$ is equivalent to the two conditions: (a) $\mathbb{E}[|X_\alpha|]$ is uniformly bounded and (b) to every $\epsilon > 0$ there corresponds a $\delta > 0$ such that $\mathbb{P}(A) < \delta$ implies $\mathbb{E}[|X_\alpha| : A] < \epsilon$ for all α. If the probability space is non-atomic, the first condition follows from the second.

Hint: Under condition (a) for large N, by Chebyshev $\mathbb{P}[|X_\alpha| > N]$ is uniformly small. Therefore, by condition (b), we get (2.1.7). If there are no atoms, for $\epsilon = 1$, we can write the space as a union of a finite number of sets each of which has probability $\leq \delta$ such that the integral of $|X_\alpha|$ on each of these sets is ≤ 1 for all α. Hence, condition (a) is automatic.

2.1.7. A martingale $\{X_n\}$ is uniformly integrable if and only if these exists X such that $X_n = \mathbb{E}[X \mid \mathcal{F}_n]$.

Hint: Let $X = \lim X_n$.

2.1.8. Let $\mathcal{F}_n$ be a sequence of σ-fields. $\mathcal{F}_n$ is said to "converge" to a σ-field $\mathcal{F}$, written $\lim \mathcal{F}_n = \mathcal{F}$ if $\cap_n \mathcal{G}_n = \mathcal{F} = \sigma$-field generated by $\cup_n \mathcal{H}_n$ where $\mathcal{G}_n = \sigma$-field generated by $\cup_{m \geq n} \mathcal{F}_m$ and $\mathcal{H}_n = \cap_{m \geq n} \mathcal{F}_m$. For integrable X, $\mathbb{E}[X \mid \mathcal{F}_n]$ converges in L^1 to $\mathbb{E}[X \mid \mathcal{F}]$.

2.1.9. Let F_n be an increasing sequence of σ-fields and X_n integrable random variables such that $[X_n] \leq \varphi$, where $\varphi \in L^1$ and $\lim X_n = X$ almost surely. Then, $\lim Y_n = \mathbb{E}[X_\infty \mid F_\infty]$, where $F_\infty = \sigma$-field generated by $\cup_n F_n$ and $Y_n = \mathbb{E}[X_n \mid F_n]$. A similar result holds for a decreasing sequence of σ-fields.

2.1.10. Let F and G be σ-fields. F and G are said to be *conditionally independent* given $F \cap G$ if for all F-measurable X and G-measurable Y, $\mathbb{E}[XY \mid F \cap G] = \mathbb{E}[X \mid F \cap G] \cdot \mathbb{E}[Y \mid F \cap G]$. Show that F and G are conditionally independent given $F \cap G$ if and only if for all z, $\mathbb{E}[\mathbb{E}[Z \mid F] \mid G] = \mathbb{E}[\mathbb{E}[Z \mid G] \mid F]$.

2.1.11. Let F_n be an increasing sequence of σ-fields. Four stop rules T, S, F_T, and F_S are conditionally independent given $F_{T \wedge S}$.

2.2 Continuous Parameter

Let $F(t)$, $0 \leq t < \infty$, be an increasing family of σ-fields. Assume also that the family is *right continuous*:

$$F(t) = \bigcap_{s>t} F(s).$$

As in the discrete case, a family $X(t)$ of integrable random variables is called a supermartingale if $X(t)$ is $F(t)$-measurable and

$$\mathbb{E}[X(t) \mid F(s)] \leq X(s) \quad s \leq t,$$

a sub-martingale if $-X(t)$ is a supermartingale and a martingale if it is both sub and super.

A supermartingale is called *right continuous* if for almost all t, $t \mapsto X(t)$ is right continuous.

A *stopping time* T is a non-negative random variable such that

$$(T \leq t) \in F(t) \quad \text{for all } t.$$

Each stopping time T is the limit of a decreasing sequence of *discrete stopping times*: Define T_n by

$$T_n = \begin{cases} (i+1)2^{-n} & i2^{-n} \leq T < (i+1)2^{-n} \\ \infty & T = \infty. \end{cases} \quad (2.2.1)$$

It is easy to check that T_n are stopping times and decrease to T.

Thus, if $X(\cdot)$ is right continuous, $X(T)$ is a random variable for any stopping time T.

For a stopping time T, the σ-field $F(T)$ is defined by

$$F(T) = \{A : A \cap (T \leq t) \in F(t) \text{ for all } t\}.$$

Then, if $T \leq S$, $F(T) \subset F(S)$, and if $X(\cdot)$ is right continuous, $X(T)$ is $F(T)$-measurable.

The right continuity of the σ-fields $F(t)$ implies that $F(T_n)$ decreases to $F(T)$ whenever the stopping times T_n decrease to T.

All the above statements are easily proved. We do not do this here because they are done for the Brownian motion in Chapter 3 and the proofs in the general case are essentially the same.

Proposition 2.2.1 (Optional Sampling Theorem). *Let $X(t)$ be a right continuous supermartingale. If $T \leq S$ are bounded stopping times, then*

$$\mathbb{E}[X(T)] \geq \mathbb{E}[X(S)]. \tag{2.2.2}$$

And there is equality in case of a martingale.

Proof. If T and S are discrete, this is just Corollary 2.1.2; (2.1.4) is simply another way of writing (2.2.2) if $S \leq N$. Now suppose T_n is as in (2.2.1) and decreases to T. By the right continuity, $X(T_n)$ tends to $X(T)$. That $X(T_n)$ is a supermartingale relative to the decreasing sequence $F(T_n)$ of σ-fields follows from (2.2.2) (for discrete stopping times) and Remark 2.1.3. Since $\mathbb{E}[X(T_n)] \leq \mathbb{E}[X(0)]$, Proposition 2.1.9 applies and we can conclude that $X(T_n)$ tends to $X(T)$ in L^1. $\square$

Corollary 2.2.2. *Let $X(t)$ be a non-negative right continuous submartingale. Then for any $a > 0$,*

$$a\mathbb{P}\left[\sup_t X(t) \geq a\right] \leq \sup_t \mathbb{E}[X(t)].$$

Proof is exactly as that of Lemma 2.1.11.

We shall use the following simple proposition in Example 2.2.5.

Proposition 2.2.3. *Let $X(t)$ be a martingale such that $|X(t)| \leq Mt$ for some constant M. If T is a stopping time such that $\mathbb{E}[T] < \infty$, then $\mathbb{E}[X(T)] = \mathbb{E}[X(0)]$.*

Proof. $T < \infty$ almost everywhere. Apply (2.2.2) with $S = 0$ and $T \wedge n$, instead of T. The variables $X(T \wedge n)$ tend to $X(T)$ and are bounded by the integrable function $M.T.$, and dominated converges can be used to conclude the proof. □

In the following example, a knowledge of relevant terms is assumed.

Example 2.2.4. This example illustrates a simple application of the martingale convergence theorem to excessive functions. We show that if f and g are excessive (see Section 6.1), so is $h = f \wedge g$.

Let X_t be the Brownian motion for any standard Markov process. We claim that if t_n is a sequence decreasing to zero, then

$$\mathbb{P}_a \left[\lim f(X_{t_n}) = f(a) \right] = 1. \tag{2.2.3}$$

Indeed, for any positive number A, $f(X_{t_n}) \wedge A$ being a non-negative bounded supermartingale relative to a decreasing sequence of σ-fields converges almost everywhere and in L^1. $\lim f(X_{t_n})$ thus exists almost everywhere, which by the zero–one law must be a constant, say B.

By L^1-convergence of $f(X_{t_n}) \wedge A$,

$$f(a) = \lim \mathbb{E}_a \left[f(X_t) \right] \geq \lim \mathbb{E}_a \left[f(X_{t_n}) \wedge A \right]$$
$$= B \wedge A \geq \mathbb{E}_a \left[f(X_t) \wedge A \right]$$

because the last quantity is a decreasing function of t. Letting A tend to infinity and t tend to zero, we get (2.2.3).

If f and g are excessive and $h = f \wedge g$, then for all t, $\mathbb{E}_a [h(X_t)] \leq h(a)$. Form (2.2.3) as t_n decreases zero $h(X_{t_n})$ tends to $h(a)$ and so an appeal to Fatou shows that h is indeed excessive.

Example 2.2.5. Let u be twice continuously differentiable in $\mathbb{R}^d$ and suppose that Δu is bounded, Δ denoting Laplacian. If X_t denotes Brownian motion, then

$$u(X_t) - u(X_0) - \frac{1}{2} \int_0^t \Delta u(X_s) \, ds \tag{2.2.4}$$

is a martingale relative to $\mathbb{P}_a$, for any $a \in \mathbb{R}^d$. To see this, denote by p the heat kernel $p(t, x) = (2\pi t)^{-d/2} \exp(-|x|^2 / 2t)$, which satisfies

$$\frac{1}{2} \Delta p = \frac{\partial}{\partial t} p.$$

Then,
$$\frac{\partial}{\partial t} u * p(t, \cdot) = u * \frac{\partial}{\partial t} p(t, \cdot)$$
$$= u * \frac{1}{2}\Delta p = \frac{1}{2}\Delta u * p,$$

which when integrated leads to

$$\mathbb{E}.\left[u\left(X_t\right)\right] - u\left(\cdot\right) = \frac{1}{2}E.\left[\int_0^t \Delta u\left(X_s\right) ds\right].$$

Equation (2.2.4) follows easily from Markov property and the last identity. Proposition 2.2.3 now implies the following:

If T is a stopping time such that $\mathbb{E}_a\left[T\right] < \infty$, then

$$\mathbb{E}_a\left[u\left(X_T\right)\right] - u\left(a\right) = \frac{1}{2}\mathbb{E}_a\left[\int_0^T \Delta u\left(X_s\right) ds\right].$$

This is known as Dynkin's formula. More on this in Chapter 5.

For more on martingales, consult the above book of Meyer [14] and Strasbourg Lecture Notes.

References

[1] J. Backhoff-Veraguas and C. Unterberger. On the specific relative entropy between martingale diffusions on the line. *Electron. Commun. Probab.*, 28: 12, 2023. Paper No. 37.
[2] E. J. Baurdoux, A. E. Kyprianou and C. Ott. Optimal prediction for positive self-similar Markov processes. *Electron. J. Probab.*, 21: 24, 2016. Paper No. 48.
[3] R. Belfadli, M. Chadad and M. Erraoui. On the sum of Gaussian martingale and an independent fractional Brownian motion. *Theory Probab. Appl.*, 68(2): 316–323, 2023. Reprint of *Teor. Veroyatn. Primen.* 6(8): 383–392, 2023.
[4] R. Bhattacharya and E. C. Waymire. *Random Walk, Brownian Motion, and Martingales*. Graduate Texts in Mathematics, Vol. 292. Springer, Cham, 2021.
[5] S. Karlin and H. M. Taylor. *A First Course in Stochastic Processes*, 2nd edn. Academic Press [Harcourt Brace Jovanovich, Publishers], New York, 1975.
[6] O. Kella. Martingales associated with functions of Markov and finite variation processes. *Queueing Syst.*, 100(3–4): 205–207, 2022.

[7] V. Knopova, A. Kulik and R. L. Schilling. Construction and heat kernel estimates of general stable-like Markov processes. *Dissertationes Math.*, 569: 86, 2021.

[8] N. E. Kordzakhia, A. A. Novikov and A. N. Shiryaev. The Kolmogorov inequality for the maximum of the sum of random variables and its martingale analogues. *Theory Probab. Appl.*, 68(3): 457–472, 2023.

[9] T. G. Kurtz and J. Swanson. Finite Markov chains coupled to general Markov processes and an application to metastability I. In *Stochastic Analysis, Filtering, and Stochastic Optimization*, pp. 293–307. Springer, Cham, 2022.

[10] A. E. Kyprianou and A. R. Watson. Potentials of stable processes. In *Séminaire de Probabilités XLVI*. Lecture Notes in Mathematics, Vol. 2123, pp. 333–343. Springer, Cham, 2014.

[11] A. E. Kyprianou. *Gerber-Shiu Risk Theory*. European Actuarial Academy (EAA) Series. Springer, Cham, 2013.

[12] A. E. Kyprianou and J. C. Pardo. *Stable Lévy Processes via Lamperti-Type Representations*. Institute of Mathematical Statistics (IMS) Monographs, Vol. 7. Cambridge University Press, Cambridge, 2022.

[13] K. Liu, J. Lu and L. Peng. Real interpolation between strong martingale Hardy spaces. *Acta Math. Vietnam.*, 48(3): 423–443, 2023.

[14] P.-A. Meyer. *Probability and Potentials*. Blaisdell Publishing Co. [Ginn and Co.], Waltham, Mass., 1966.

[15] Y. Miura. The conservativeness of Girsanov transformed symmetric Markov processes. *Tohoku Math. J. (2)*, 71(2): 221–241, 2019.

[16] M. M. Rao. *Conditional Measures and Applications*. Monographs and Textbooks in Pure and Applied Mathematics, Vol. 177. Marcel Dekker, Inc., New York, 1993.

[17] M. M. Rao. *Stochastic Processes: General Theory*. Mathematics and Its Applications, Vol. 342. Kluwer Academic Publishers, Dordrecht, 1995.

[18] M. M. Rao. Exploring ramifications of the equation $E(Y \mid X) = X$. *J. Stat. Theory Pract.*, 1(1): 73–88, 2007.

[19] M. M. Rao and R. J. Swift. From additive to second-order processes. In *Stochastic Processes and Functional Analysis — New Perspectives*. Contemporary Mathematics, Vol. 774, pp. 205–215. American Mathematical Society, Providence, 2021.

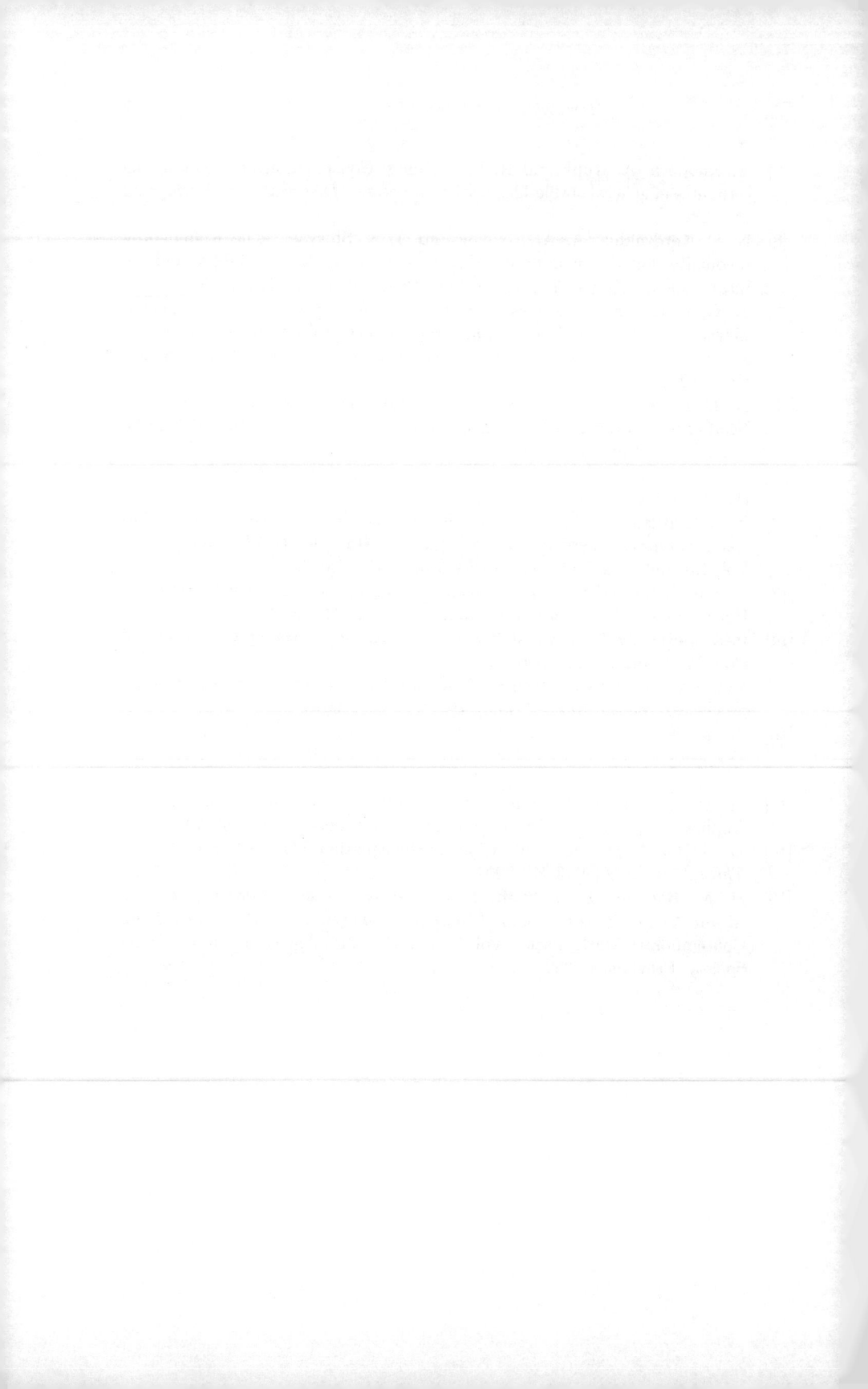

Chapter 3

Brownian Motion and Ito Calculus

Brownian motion was first introduced as a Gaussian process indexed by one-dimensional time by L. Bachelier, A. Einstein, and N. Wiener. Paul Lévy introduced the notion of quadratic variation, and he proved that every continuous martingale with quadratic variation equal to time is a copy of Brownian motion. Bachelier is considered both the forefather of mathematical finance and a pioneer in the study of stochastic processes. The purpose of this chapter is an extension from 1D to the multivariable Brownian motion.

The original material in Chapter 2 dealing with Brownian motion is still current, and we have added a discussion of new directions. In addition, for the benefit of readers, we offer the following citations covering new directions [1–3, 5–7, 9, 11–17].

The new directions include Karhunen–Loève expansions, transfer operators and conditional expectations, upcrossing chains of stopped Brownian motion, Brownian motion ensembles of random matrix theory, fractal Brownian motion, and quantum fields from Markoff fields.

For the benefit of beginner readers, we offer the following supplementary introductory text [8].

Introduction and summary. In this chapter, we introduce the d-dimensional Brownian motion, assuming as the known one-dimensional Brownian motion starting at 0. Then, we consider Markov times in more detail than is absolutely necessary, in the hope that this will better the understanding of this important notion. Strong Markov property is proved and a few applications considered.

3.1 The d-Dimensional Brownian Motion

As proved, for instance, in [4, pp. 12–16], there is a real-valued stochastic process $\xi(t)$, $0 \le t < \infty$ (on some probability space), such that $\xi(t)$ is continuous, $\xi(0) \equiv 0$, and with the finite dimensional distributions:

$$\mathbb{P}\left[\xi(t_i) \in E_i, 1 \le i \le n\right]$$
$$= \int_{E_1} p'(t_1, 0, da_1) \int_{E_2} p'(t_2 - t_1, a_1, da_2) \cdots \int_{E_n} p'(t_n - t_{n-1}, a_{n-1}, da_n),$$
(3.1.1)

where $0 < t_1 < \cdots < t_n$, $p'(t, a, db) = \frac{1}{\sqrt{2\pi t}} \exp\left(-\frac{(b-a)^2}{2t}\right) db$ and $E_1, \ldots, E_n$ are Borel subsets of the real line. See also pp. 5–8 of [10] for another proof of this fact.

Now, let us define the d-dimensional Brownian motion. There is a probability space $(\Omega, \mathscr{B}, \mathbb{P})$ and a d-dimensional stochastic process $X(t) = (X_1(t), \ldots, X_d(t))$ such that $X_1(\cdot), \ldots, X_d(\cdot)$ are independent stochastic processes and each is a copy of $\xi(\cdot)$. Let W be the space of continuous paths $t \to W(t) \in \mathbb{R}^d$, $t \ge 0$. In W, consider the smallest Borel field $\mathscr{B}_\infty$ which makes all the coordinate maps measurable: $\mathscr{B}_\infty$ is the smallest Borel field relative to which all the maps $t \to W(t)$ are measurable for all $t \ge 0$. The map

$$\Omega \ni w \xrightarrow{\varphi_a} x_t^a(w) = a + X_t(w) \in W, \quad a \in \mathbb{R}^d,$$

is clearly measurable and induces probability $\mathbb{P}_a$ on $(W, \mathscr{B}_\infty)$.

$(W, \mathscr{B}_\infty, \mathbb{P}_a, a \in \mathbb{R}^d, x_t)$ is called the *standard path space realization* of the Brownian motion or, simply, the standard d-dimensional Brownian motion, where x_t is now the coordinate mapping of the path space W:

$$x_t(w) = x(t, w) = w(t).$$

The standard Brownian motion is a collection of individual stochastic processes, each starting at a point $a \in \mathbb{R}^d$ knitted together in a certain manner: This is the so-called (simple) *Markov property*:

$$\mathbb{P}_a\left[\bigcap_{k \le m}(x(t_k) \in E_k)\right]$$
$$= \int_{E_1} \cdots \int_{E_m} p(t_1, a, dx_1) p(t_2 - t_1, x_1, dx_2) \cdots p(t_m - t_{m-1}, x_{m-1}, dx_m),$$
(3.1.2)

where $E_1, \ldots, E_m$ are Borel subsets of $\mathbb{R}^d$ and $0 < t_1 < t_2 < \cdots < t_m$, $m \geq 1$, and $p(t, a, db) = \mathbb{P}_a[x(t) \in db]$. We also have

$$\mathbb{P}_a(B) = \mathbb{P}_0(w + a \in B),$$
$$\mathbb{P}_a(-w \in B) = \mathbb{P}_{-a}(B), \qquad B \in \mathscr{B}_\infty, \qquad (3.1.3)$$

$$\mathbb{P}_a(x(0) = a) = 1. \qquad (3.1.4)$$

$\mathbb{P}_a(B)$ is to be thought of as *the chance that the event $B \in \mathscr{B}_\infty$ occurs for the Brownian path starting at $a \in \mathbb{R}^d$*. Equation (3.1.2) is easy to verify using (3.1.1).

In general, an event $B \in \mathscr{B}_\infty$ depends on an infinite set of parameter values t. However, certain statements which are true for all events depending on finitely many parameters also hold for all events in $\mathscr{B}_\infty$. We counter several examples in the sequel. The following theorem will be found useful in reaching such a conclusion.

Theorem 3.1.1. *Let H be a vector space of bounded real-valued functions defined on a set X, which contains the constant 1, is closed under uniform convergence, and is such that for every increasing uniformly bounded sequence f_n of non-negative functions $f_n \in H$, the function $f = \lim f_n \in H$. Let C be a subset of H, closed under multiplication. Then the space H contains all bounded functions measurable with respect to the Borel field generated by the elements of C.*

Proof. If $f_1, \ldots, f_n \in C$, any polynomial in $f_1, \ldots, f_n$ belongs to H. The conditions on H together with Stone–Weierstrass theorem imply that for any continuous function φ on $\mathbb{R}^n$, $\varphi(f_1, \ldots, f_n) \in H$. H is closed under uniformly bounded increasing limits; since H contains constants, it is closed under uniformly bounded decreasing limits. The set of φ on $\mathbb{R}^n$ for which $\varphi(f_1, \ldots, f_n) \in H$ contains the bounded continuous functions and is closed under uniformly bounded monotone limits. It therefore contains all bounded Borel measurable functions. We leave the rest of the proof to the reader. $\square$

As the first example of the way Theorem 3.1.1 is used, let us show that $\mathbb{E}_a[f]$ is a-measurable on $\mathbb{R}^d$ for each bounded $\mathscr{B}_\infty$-measurable function f. The set of such functions clearly satisfies the conditions of Theorem 3.1.1 and contains the multiplicative class of $\mathscr{B}_\infty$-measurable bounded functions depending on finitely many parameters, i.e., functions f of the form

$f(w) = \varphi(x_{t_1}(w), \ldots, x_{t_n}(w))$, where φ is bounded and measurable on $\mathbb{R}^{nd}$. By Theorem 3.1.1, it contains all $\mathscr{B}_\infty$-measurable bounded functions.

The reader will note that sometimes we need to conclude the joint measurability of some stochastic processes. The following theorem will be found useful.

Theorem 3.1.2. *Let $X(t, w)$ be a stochastic process, $t \in [0, \infty)$. If $X(t, w)$ is measurable in w for each t and is right continuous in t for each w, then $X(t, w)$ is measurable in the pair (t, w).*

Proof. The sequence $X_n(t, w)$ of stochastic processes defined by

$$X_n(t, w) = X\left(\frac{i+1}{2^n}, w\right), \quad \frac{i}{2^n} \leq i < \frac{i+1}{2^n}, \quad i = 0, 1, 2, \ldots$$

for $n = 1, 2, \ldots$ is jointly measurable, and

$$\lim_n X_n(t, w) = X(t, w). \qquad \square$$

With the help of (3.1.2), it is seen that

$$\mathbb{P}_a\left[x(t) \in A; \bigcap_{k \leq m} (x(t_k) \in E_k)\right]$$

$$= \mathbb{E}_a\left[\int_A p(t - t_m, x(t_m), b)\, db : \bigcap_{k < m} (x(t_k) \in E_k)\right]$$

$$= \mathbb{E}_a\left[\mathbb{P}_{x(t_m)}\left[x(t - t_m) \in A : \bigcap_{k \leq m} (x(t_k) \in E_k)\right]\right] \quad (3.1.5)$$

for $t_1 < t_2 < \cdots < t_m < t$.

We deduce

$$\mathbb{P}_a[x(t+s) \in A \mid \mathscr{B}_s] = \mathbb{P}_{x(s)}(x(t) \in A), \quad (3.1.6)$$

where $\mathscr{B}_s = \mathscr{B}[x(t_1); t_1 \leq s]$ is the smallest Borel field generated by the variables $x(t_1)$, $t_1 \leq s$.

That (3.1.5) implies (3.1.6) is an example of the use of Theorem 3.1.1.

Markov property

The Markov property has several different expressions which are all equivalent but each has some technical advantage. We are concerned with the standard dimensional Brownian motion. Before discussing Markov property, let us introduce maps $W \to W$:

- the *shift operators* θ_t ($t \geq 0$) defined by $\theta_t w(s) = w(s+t)$, $s \leq 0$;
- the *stopping operator* α_t ($0 \leq t \leq \infty$) defined by $\alpha_t w(s) = w(s \wedge t)$, $s \geq 0$, so that $\alpha_\infty w = w$.

It is easy to verify that both are measurable maps of $(W, \mathscr{B}_\infty)$ into $(W, \mathscr{B}_\infty)$. We leave it to the reader to verify the following:

Fact. *The smallest Borel field $\mathscr{B}_t$ relative to which α_t is measurable is precisely the Borel field of events depending on the sample path to time t, i.e., $\mathscr{B}_t = \mathscr{B}(x_s : s \leq t) =$ the least Borel field relative to which all maps $w \to w(s)$ are measurable for all $s \leq t$.*

The standard Brownian motion has the following Markov property:

$$\mathbb{P}_a\left[\theta_s^{-1} A \cap B\right] = \mathbb{E}_a\left[\mathbb{P}_{x_s}(A); B\right], \quad A \in \mathscr{B}_\infty,\ B \in \mathscr{B}_s,$$
$$\mathbb{P}_a\left[\theta_s^{-1} A \mid \mathscr{B}_s\right] = \mathbb{P}_{x_s}(A), \qquad A \in \mathscr{B}_\infty, \qquad (3.1.7)$$

$$\mathbb{E}_a\left[G \circ \theta_t \cdot F\right] = \mathbb{E}_a\left[\mathbb{E}_{x_t}(G) \cdot F\right],$$
$$\mathbb{E}_a\left[G \circ \theta_t \mid \mathscr{B}_t\right] = \mathbb{E}_{x_t}(G) \qquad (3.1.8)$$

for any G, F respectively bounded $\mathscr{B}_\infty$- and $\mathscr{B}_t$-measurable functions:

$$\mathbb{E}_a\left[f(x_t(\theta_s w)) \mid \mathscr{B}_s\right] = \mathbb{E}_{x_s}(f(x_t)) \qquad (3.1.9)$$

for any bounded Borel-measurable function f on $\mathbb{R}^d$.

That (3.1.7) and (3.1.8) are equivalent is general measure theory. Equation (3.1.9) is a particular case of (3.1.8), and (3.1.9) is equivalent to equation (3.1.6). Let us show that (3.1.9) implies (3.1.8). Use induction and suppose G has the form

$$G(w) = f_1(x_{t_1}(w)) \cdots f_m(x_{t_m}(w)),$$

where $t_1 < t_2 < \cdots < t_m$ and $f_1, \ldots, f_m$ are bounded Borel measurable functions on $\mathbb{R}^d$. $G \circ \theta_t(w) = f_1(x_{t_1+t}(w)) \cdots f_m(x_{t_m+t}(w))$. And

$$\mathbb{E}_a \left[\prod_1^m f_i(x_{t_i+t}) \mid \mathscr{B}_t \right] = \mathbb{E}_a \left[\mathbb{E}_a \left[\prod_1^m f_i(x_{t_i+t}) \mid \mathscr{B}_{t_{m-1}} \right] \mid \mathscr{B}_t \right]$$

$$= \mathbb{E}_a \left[\prod_1^{m-1} f_i(x_{t_i+t}) \, \mathbb{E}_a \left[f_m(x_{t_m+t}) \mid \mathscr{B}_{t_{m-1}+t} \right] \mid \mathscr{B}_t \right]$$

$$\left(\text{since } \prod_1^{m-1} f_i(x_{t_i+t}) \text{ is } \mathscr{B}_{t_{m-1}+t}\text{-measurable} \right)$$

$$= \mathbb{E}_a \left[\prod_1^{m-1} f_i(x_{t_i+t}) \, \mathbb{E}_{x_{t+t_{m-1}}} \left(f_m(x_{t_m-t_{m-1}}) \right) \mid \mathscr{B}_t \right]$$

$$\left(\text{since } \mathbb{E}_a \left[f_m(x_{t_m+t}) \mid \mathscr{B}_{t_{m-1}+t} \right] = \mathbb{E}_{x_{t+t_{m-1}}} \left(f_m(x_{t_m-t_{m-1}}) \right) \right)$$

$$= \mathbb{E}_{x_t} \left[\prod_1^{m-1} f_i(x_{t_i}) \, \mathbb{E}_{x_{t_{m-1}}} \left(f_m(x_{t_m-t_{m-1}}) \right) \right]$$

$$\left(\text{by induction assumption applied to } G_1, \text{ where} \right.$$

$$\left. G_1(w) = f_1(x_{t_1}(w)) \cdots f_{m-1}(x_{t_1}(w)) \, \mathbb{E}_{x_{t_{m-1}}(w)} \left(f_m(x_{t_m-t_{m-1}}) \right) \right)$$

$$= \mathbb{E}_{x_t} \left[\prod_1^m f_i(x_{t_i}) \right]$$

(apply (3.1.9) with $s = t_{m-1}$, $t = t_m - t_{m-1}$, and $a = x_t(w)$). Now, Theorem 3.1.1 takes over.

3.2 Strong Markov Property

The Brownian motion also starts afresh at certain *random* times, such as the hitting time

$$T_U = \inf(t > 0 : x(t) \in U),$$

where U is an open set in $\mathbb{R}^d$, and $T_U = \infty$ if $x(t)$ never hits U, instead of a constant time t:

$$\mathbb{E}_a \left[f \left(x_s \left(\theta_T w \right) \right) \mid \mathscr{B}_T \right] = \mathbb{E}_{x_T} \left[f \left(x_s \right) \right].$$

This property was familiar and extensively used to derive deep results by, for example, P. Lévy. The complete statement of this feature of the Brownian motion, however, was discovered by Hunt and Dynkin independently in the early 1950s.

A random variable $T : W \to [0, \infty]$ is said to be a *Markov time* (or *stopping time*) if

$$\{ w : T(w) < t \} \in \mathscr{B}_t, \quad t \geq 0.$$

The hitting time T_U is a Markov time since

$$(T_U < t) = \bigcup_{r \in \mathbb{Q}, r < t} (x_r \in U) \in \mathscr{B}_t.$$

A constant time $T \equiv t$ is trivially a Markov time, but a last exit time such as $\sup (t \leq 1 : x(t) = 0)$ is not.

Define $\mathscr{B}_{T+}$ to be the class of sets $B \in \mathscr{B}_\infty$ such that

$$B \cap (T < t) \in \mathscr{B}_t, \quad t \geq 0.$$

$\mathscr{B}_{T+}$ is a Borel algebra and $(T < t) \in \mathscr{B}_{T+}$ for each $t > 0$. $\mathscr{B}_{T+}$ is to be thought of as measuring the Brownian path up to time $t = T+$ because

$$\mathscr{B}_{T+} = \bigcap_{\epsilon > 0} \mathscr{B} \left[x \left(t \wedge (T + \epsilon) \right) : t \geq 0 \right].$$

We shall show this a little later.

Dynkin–Hunt's statement of the *strong Markov property* is that, conditionally on the present position $x(T)$, the future path $x(t+T)$, $t \geq 0$, is a standard Brownian motion starting at $x(T)$, and this Brownian motion is independent of $\mathscr{B}_{T+}$.

Before going into precise mathematical statements, let us note some facts. By Theorem 3.1.2, $x(t, w)$ is measurable in the pair (t, w). Hence, for any non-negative measurable function b on W, the function $x(b(w), w)$ is measurable. In particular, for a Markov time T, x_T is measurable. We already defined the shift operators θ_t and stopping operators α_t for all $t \geq 0$. θ_T and α_T make sense (if $T < \infty$) and are easily seen to be measurable maps from W into W.

Theorem 3.2.1. *The Brownian motion has the strong Markov property:*

$$\mathbb{E}_a \left[G \circ \theta_T : B \cap (T < \infty) \right] = \mathbb{E}_a \left[\mathbb{E}_{x(T)}(G) : B \cap (T < \infty) \right] \quad (3.2.1)$$

for all bounded $\mathscr{B}_\infty$-measurable functions G and all $B \in \mathscr{B}_{T+}$.

Proof. We need only prove (3.2.1) for functions G of the form

$$G(w) = f_1(x(t_1)) \cdots f_m(x(t_m)),$$

where $f_1, \ldots, f_m$ are continuous bounded functions on $\mathbb{R}^d$. Once we do this, Theorem 3.1.1 takes over.

Define a sequence (of stopping times) T_n as follows:

$$T_n(w) = \begin{cases} k2^{-n} & (k-1)2^{-n} \leq T < k2^{-n},\ k \geq 1, \\ \infty & T = \infty. \end{cases}$$

Noting the following facts (we leave the proofs to the reader):

1. $G(\theta_T w) = \lim G(\theta_{T_n} w)$.
2. $\mathbb{E}_a[G]$ is a continuous function of a.
3. $B \cap (T_n = k2^{-n}) = B \cap ((k-1)2^{-n} \leq T < k2^{-n}) \in \mathscr{B}_{k2^{-n}}$, $k \geq 1$, for all $B \in \mathscr{B}_{T+}$.

We write

$$\mathbb{E}_a \left[G \circ \theta_T : B \cap (T < \infty) \right] = \lim_{n \to \infty} \mathbb{E}_a \left[G \circ \theta_{T_n} : B \cap (T_n < \infty) \right]$$

$$= \lim_{n \to \infty} \sum_{k=1}^{\infty} \mathbb{E}_a \left[G \circ \theta_{k2^{-n}} : B \cap (T_n = k2^{-n}) \right]$$

$$= \lim_{n \to \infty} \sum_{k=1}^{\infty} \mathbb{E}_a \left[\mathbb{E}_{x(k2^{-n})}(G) : B \cap (T_n = k2^{-n}) \right]$$

(this is the simple Markov property)

$$= \lim_{n \to \infty} \mathbb{E}_a \left[\mathbb{E}_{x(T_n)}(G) : B \cap (T < \infty) \right]$$

$$= \mathbb{E}_a \left[\mathbb{E}_{x(T)}(G) : B \cap (T < \infty) \right].$$

This proves the theorem. □

Equation (3.2.1) is equivalent to

$$\mathbb{E}_a \left[G \circ \theta_T \mid \mathscr{B}_{T+} \right] = \mathbb{E}_{x(T)}(G) \qquad (3.2.2)$$

$\mathbb{P}_a$-almost everywhere on the set $(T < \infty)$.

Proceeding as in Exercise 3.2.3, we can show that

$$\mathbb{P}_a \left[\theta_T^{-1}(A) \cap B \mid x(T) \right] = \mathbb{P}_a \left[\theta_T^{-1} A \mid x(T) \right] \mathbb{P}_a \left(B \mid x(T) \right),$$

for $A \in \mathscr{B}_\infty$, $B \in \mathscr{B}_{T+}$, almost everywhere on the set $(T < \infty)$. Stated otherwise, this means that conditional on $x(T)$, the motion $x(T+t)$ is independent of $\mathscr{B}_{T+}$.

Markov times

Recall the definition of the stopping operator α_t:

$$\alpha_t w(s) = w(t \wedge s), \quad t, s \geq 0.$$

α_t has the following important property:

$$\alpha_t \left[\alpha_t w \right] = \alpha_t w. \qquad (3.2.3)$$

The reader will presently see the use of the following proposition in the study of Markov times.

Proposition 3.2.2. $B \in \mathscr{B}_t$ if and only if

$$B = \alpha_t^{-1}(B).$$

Proof. This is immediate: Since $\mathscr{B}_t$ is the smallest Borel field relative to which α_t is measurable, there exists a set $A \in \mathscr{B}_\infty$ such that

$$B = \alpha_t^{-1}(A).$$

We have $\alpha_t^{-1}(B) = \alpha_t^{-1}\alpha_t^{-1}(A) = \alpha_t^{-1}(A) = B$ from (3.2.3). □

Corollary 3.2.3. Let $\{B_i : i \in I\}$ be a collection of sets in $\mathscr{B}_t$. If $B = \cup_{i \in I} B_i \in \mathscr{B}_\infty$, then $B \in \mathscr{B}_t$.

Indeed, from the above proposition,

$$\alpha_t^{-1}(B) = \bigcup_i \alpha_t^{-1}(B_i) = \bigcup_i B_i = B.$$

Introduce the Borel fields $\mathscr{B}_{t+}$:

$$\mathscr{B}_{t+} = \bigcap_{s>t} \mathscr{B}_s, \quad t \geq 0.$$

An immediate corollary of Corollary 3.2.3 is as follows:

Corollary 3.2.4. *Let $\{B_i : i \in I\}$ be a collection (not necessarily countable) of sets in $\mathscr{B}_{t+}$. If $B = \cup_i B_i \in \mathscr{B}$, then $B \in \mathscr{B}_{t+}$.*

Our definition of a Markov time can be rewritten as follows:

$$(T \leq t) \in \mathscr{B}_{t+}, \quad t \geq 0.$$

Proposition 3.2.5. *Let $T : W \to [0, \infty]$ be $\mathscr{B}_\infty$-measurable. Then T is a Markov time if and only if*

$$(T = t) \in \mathscr{B}_{t+}, \quad t \geq 0.$$

Proof. Indeed, $(T \leq t) = \cup_{s \leq t}(T = s)$ and Corollary 3.2.4 takes over. □

Using Proposition 3.2.5, it is a simple matter *to check that the class of Markov times is closed under the following operations*:

$$T_1 \wedge T_2, \qquad T_1 \vee T_2,$$
$$T_n \downarrow T, \qquad T_n \uparrow T,$$
$$T_1 + T_2,$$
$$T_1 + T_2(\theta_{T_1} w).$$

Consider, as an example, the proof that $S = T_1 + T_2(\theta_{T_1} w)$ is a Markov time whenever T_1, T_2 are. First, S is $\mathscr{B}_\infty$-measurable. And

$$(S = t) = \bigcup_{\substack{r,s \\ r+s=t}} (T_1 = r) \cap (T_2(\theta_r) = s).$$

Now, note that for all r, s, $\theta_r^{-1}(A) \in \mathscr{B}_{(r+s)+}$ for all $A \in \mathscr{B}_{S+}$ (first prove that $\theta_r^{-1}(A) \in \mathscr{B}_{(r+s)+}$ for all $r, s \geq 0$). Thus, $(S = t) \in \mathscr{B}_{t+}$ and the proof is complete.

Before examining Markov times further, let us introduce strict stopping times:

A non-negative function $T \leq \infty$ is a *strict Markov time* if for all $t \geq 0$,

$$(T \leq t) \in \mathscr{B}_t.$$

Clearly, every strict Markov time is also a Markov time; if T is a Markov time, $T + \epsilon$ is a strict Markov time for all $\epsilon > 0$. In particular, every Markov time is a limit of a decreasing sequence of strict Markov times.

We can easily show (cf. Proposition 3.2.5) that a *non-negative measurable function $T \leq \infty$ is a strict Markov time if and only if* $(T = t) \in \mathscr{B}_t$ for all $t \geq 0$.

For a strict Markov time T, the *Borel field* $\mathscr{B}_T$ is defined as the Borel field of all sets $B \in \mathscr{B}_\infty$ such that

$$B \cap (T \leq t) \in \mathscr{B}_t \quad \text{for all } t \geq 0.$$

If $T \leq S$ are (strict) Markov times and $B \in \mathscr{B}_{T+}$ (resp. $B \in \mathscr{B}_T$), we have

$$B \cap (S < t) = B \cap (T < t) \cap (S < t) \in \mathscr{B}_t$$

resp.

$$B \cap (S \leq t) = B \cap (T \leq t) \cap (S \leq t) \in \mathscr{B}_t$$

for all $t \geq 0$, i.e., $B \in \mathscr{B}_{S+}$ (resp. $B \in \mathscr{B}_S$). Thus, $\mathscr{B}_{T+} \subset \mathscr{B}_{S+}$ ($\mathscr{B}_T \subset \mathscr{B}_S$). Using this, it is simple to show that if a sequence of Markov times T_n decreases to a Markov time T,

$$\mathscr{B}_{T+} = \bigcap_n \mathscr{B}_{T_n}$$

and

$$\mathscr{B}_{T+} = \bigcap_{\epsilon > 0} \mathscr{B}_{T+\epsilon}.$$

(Note that $T + \epsilon$ is a strict Markov time for all $\epsilon > 0$.)

Suppose T is a Markov time. Then,

$$(T = t) \in \mathscr{B}_{t+} \subset \mathscr{B}_s \quad \text{for all } s > t,$$
$$\alpha_s^{-1}(T = t) = (T = t) \quad \text{for all } s > t,$$
$$T(w) = t \quad \text{implies } T(\alpha_s w) = t \quad \text{for all } s > t.$$

Thus,

$$T(w) = T(\alpha_{T+\epsilon} w) \quad \text{for all } \epsilon > 0. \tag{3.2.4}$$

Similarly, for a strict Markov time T we can verify

$$T(w) = T(\alpha_T w). \tag{3.2.5}$$

A little more careful analysis leads to Galmarino's characterization of Markov and strict Markov times.

Theorem 3.2.6 (Galmarino's Theorem). *A non-negative Borel function $T \leq \infty$ is a Markov time (a strict Markov time) if and only if*

$$\alpha_t w = \alpha_t v, \ Tw < t \text{ implies } Tw = Tv.$$
$$(\alpha_t w = \alpha_t v, \ Tw \leq t \text{ implies } Tw = Tv.)$$

Proof. If $A \in \mathscr{B}_t$, $w \in A$, $\alpha_t w = \alpha_t v$ imply $v \in A$. (This is because $A = \alpha_t^{-1}(A)$.) Thus, if T is a Markov time, $(T < t) \in \mathscr{B}_t$. Therefore, $T(w) < t$, $\alpha_t w = \alpha_t v$ imply $T(v) < t$. Hence, $T(v) = T(\alpha_t v) = T(\alpha_t w) = T(w)$.

Conversely, suppose T has the above property. Then, $(T < t) \in \mathscr{B}_\infty$ and $\alpha_t^{-1}(T < t) = (T < t)$, i.e., $(T < t) \in \mathscr{B}_t$. □

Galmarino's theorem gives us a nice intuitive idea of what a Markov time really is.

We now look more closely at the Borel fields $\mathscr{B}_{T+}$ and $\mathscr{B}_T$ (for strict Markov times). First, let T be a strict Markov time. By definition, $B \in \mathscr{B}_T$ if and only if

$$B \cap (T \leq t) \in \mathscr{B}_t$$

for all $t \geq 0$. This can also be defined as
$$B \in \mathscr{B}_\infty$$
and
$$B \cap (T = s) \in \mathscr{B}_s$$
for all $s \geq 0$. Now, we claim that $B \in \mathscr{B}_T$ if and only if $B \in \mathscr{B}_\infty$ and
$$B = \alpha_T^{-1}(B). \tag{3.2.6}$$
If $B = \alpha_T^{-1}(B)$ and $B \in \mathscr{B}_\infty$, we get
$$B \cap (T = s) = \alpha_T^{-1}(B) \cap (T = s) = \alpha_s^{-1}(B) \cap (T = s) \in \mathscr{B}_s$$
since $\alpha_T^{-1}(B) \in \mathscr{B}_s$, and $(T = s) \in \mathscr{B}_s$. Thus, $B \in \mathscr{B}_T$. Conversely, if $B \in \mathscr{B}_T$, for all s
$$B \cap (T = s) \in \mathscr{B}_s,$$
i.e.,
$$\alpha_T^{-1}(B) \cap (T = s) = B \cap (T = s) \left[\alpha_T^{-1}(T = s) = (T = s) \right].$$
Taking union over all s, we get
$$\bigcup_s \left(\alpha_T^{-1}(B) \cap (T = s) \right) = B,$$
i.e.,
$$\alpha_T^{-1}(B) = B.$$
For any $A \in \mathscr{B}_\infty$, $B = \alpha_T^{-1}(A)$ has the property $\alpha_T^{-1}(B) = B$. We thus see that $\mathscr{B}_T$ is the smallest Borel field relative to which α_T is measurable, i.e.,
$$\mathscr{B}_T = \mathscr{B}\left(x \left(s \wedge T \right), s \geq 0 \right).$$
If T is a Markov time, $T + \epsilon$ is a strict Markov time for all $\epsilon > 0$ and $\mathscr{B}_{T+} = \cap_{\epsilon > 0} \mathscr{B}_{T+\epsilon}$. Thus,
$$B \in \mathscr{B}_{T+} \quad \text{iff} \quad B \in \mathscr{B}_\infty \quad \text{and} \quad B = \alpha_{T+\epsilon}^{-1}(B) \quad \text{for all } \epsilon > 0. \tag{3.2.7}$$
And
$$\mathscr{B}_{T+} = \bigcap_{\epsilon > 0} \mathscr{B}\left(x \left(s \wedge (T + \epsilon) \right) \right), s > 0).$$
If $\mathscr{A}_1$ and $\mathscr{A}_2$ are Borel fields, we denote by $\mathscr{A}_1 \vee \mathscr{A}_2$ the least Borel field containing $\mathscr{A}_1$ and $\mathscr{A}_2$.

Proposition 3.2.7. *If T is a Markov time,*
$$\mathscr{B}_\infty = \mathscr{B}_{T+} \vee \mathscr{B}\left[1_{(T<\infty)}x(t+T), t \geq 0\right].$$
If T is a strict Markov time,
$$\mathscr{B}_\infty = \mathscr{B}_T \vee \mathscr{B}\left[1_{(T<\infty)}x(t+T), t \geq 0\right],$$
where
$$1_{(T<\infty)} = \begin{cases} 1 & T < \infty \\ 0 & T = \infty. \end{cases}$$

Proof. We need only show that for every t, $x(t)$ is measurable relative to $\mathscr{B}_{T+} \vee \mathscr{B}\left[1_{(T<\infty)}x(t+T), t \geq 0\right]$:
$$x(t) = x(t)1_{(T\geq t)} + x(t)1_{(T<t)}.$$

We can rewrite (3.2.7) in the following form: A $\mathscr{B}_\infty$-measurable function f is $\mathscr{B}_{T+}$-measurable if and only if
$$f(w) = f(\alpha_{T+\epsilon}w) \quad \text{for all } \epsilon > 0.$$

We then see that $x(t)1_{(T\geq t)}$ is $\mathscr{B}_{T+}$-measurable (use (3.2.4)). And
$$x(t)1_{(T<t)} = \lim_{n\to\infty} \sum_{k2^{-n}<t} x\left(t - (k-1)2^{-n} + T\right) 1_{((k-1)2^{-n}<T\leq k2^{-n})}.$$

The indicator function in the above sum $\mathscr{B}_{T+}$-measurable and
$$1_{(T\leq k2^{-n})}x\left(t - (k-1)2^{-n} + T\right)$$
is $\mathscr{B}\left[1_{(T<\infty)}x(t+T), t \geq 0\right]$-measurable, and we are done. □

One important consequence of the strong Markov property is as follows:

Proposition 3.2.8. $\mathscr{B}_{0+}$ *is trivial, i.e.,* $A \in \mathscr{B}_{0+}$ *implies*
$$\mathbb{P}_a(A) = 0 \quad \text{or} \quad 1, \quad \forall a \in \mathbb{R}^d.$$

Proof. Indeed,
$$\mathbb{P}_a[A] = \mathbb{P}_a[A; A] = \mathbb{P}_a\left[\theta_0^{-1}(A); A\right] = \mathbb{E}_a\left[\mathbb{P}_{x(0)}A; A\right] = \mathbb{P}_a(A) \cdot \mathbb{P}_a(A)$$
since θ_0 is the identity and $\mathbb{P}_a[x(0) = 1] = 1$.

Since for any Markov time T, the set $(T=0) \in \mathscr{B}_{0+}$, we see that $\mathbb{P}_a(T=0) = 1$ or 0. □

More generally, we can show that for any $t \geq 0$, $\mathscr{B}_t$ and $\mathscr{B}_{t+}$ are equivalent: Every set in $\mathscr{B}_{t+}$ differs from a set in $\mathscr{B}_t$ at most in a set of measure zero, i.e., the completion of $\mathscr{B}_t$ and $\mathscr{B}_{t+}$ with respect to $\mathbb{P}_a$ for any a are identical. To see this, we use Proposition 3.2.7. This proposition implies that the set of functions of the form

$$f(\alpha_t w) g(\theta_t w),$$

with f, g bounded $\mathscr{B}_\infty$-measurable functions, generates the Borel field $\mathscr{B}_\infty$. We have

$$\mathbb{E}_a[f(\alpha_t) g(\theta_t) \mid \mathscr{B}_{t+}] = f(\alpha_t) \mathbb{E}_{x(t)}(g) = \mathbb{E}_a[f(\alpha_t) g(\theta_t) \mid \mathscr{B}_t].$$

The validity of this for all f, g implies

$$\mathbb{E}_a[F \mid \mathscr{B}_{t+}] = \mathbb{E}_a[F \mid \mathscr{B}_t]$$

for all bounded $\mathscr{B}_\infty$-measurable functions F, i.e., $\mathscr{B}_{t+}$ and $\mathscr{B}_t$ are equivalent.

More generally, for a strict Markov time T, $\mathscr{B}_T$ and $\mathscr{B}_{T+}$ are equivalent, where $\mathscr{B}_{T+}$ is the intersection of the Borel fields $\mathscr{B}_{T+1/n}$. We leave the proof of this as an exercise.

We have the following simple extension of the strong Markov property — *the time-dependent strong Markov property*:

Theorem 3.2.9. *Let $F(s, w)$ be bounded and measurable in (s, w). Then,*

$$\mathbb{E}[F(T, \theta_T) \mid \mathscr{B}_{T+}] = \mathbb{E}_{x(T)}[F(s, w)]_{s=T} \quad \text{on the set } (T < \infty).$$

Proof. If $F(s, w)$ has the form $f(s) g(w)$, the above is a consequence of the strong Markov property. Now, one uses the usual procedures. □

Let us look at a particular case of the generalized strong Markov property. Let T be a Markov time and

$$\mu_a(ds, db) = \mathbb{P}_a(T \in ds, X_T \in db),$$

i.e., μ_a is the joint distribution of (T, x_T). Then,

$$\mathbb{P}_a[x_t \in E] = \mathbb{P}_a[x_t \in E, T > t] + \int_{[0,t] \times \mathbb{R}^d} \mathbb{P}_b(x_{t-s} \in E) \mu_a(dsdb). \quad (3.2.8)$$

When T is the first passage time, this is known as the *"first passage time relation"*. For the proof we let

$$F(s, t) = 1_{[0,t]}(s) 1_E(x_{t-s}(w)).$$

Then, $F(T, \theta_T w) = 1_{[0,t]}(T(w)) 1_E(x_t(w))$ so that

$$\begin{aligned}
\mathbb{P}_a[x_t \in E, T \leq t] &= \mathbb{E}_a[F(T, \theta_T)] \\
&= \mathbb{E}_a[\mathbb{E}_{x_T}[F(s, w)_{s=T}]] \\
&= \int_{[0,t] \times \mathbb{R}^d} \mathbb{P}_b(x_{t-s} \in E) \mu_a(dsdb),
\end{aligned}$$

which is what we set out to show. Let us look at the applications.

Applications

Consider the one-dimensional Brownian motion, $a > 0$, and E a Borel subset of $(0, \infty)$ and define the Markov time T by

$$T = \inf\{t : x_t = 0\},$$

where $T = \infty$ if there is no such t. Then, we have

$$\mathbb{P}_a(x_t \in E, T > t) = \int_E (p(t, a, b) - p(t, a, -b)) \, db,$$

where

$$p(t, x, y) = \frac{1}{\sqrt{2\pi t}} e^{-(x-y)^2/2t}.$$

Since $x_T = 0$ if $T < \infty$, we get from the first passage time relation

$$\mathbb{P}_a(x_t \in E) = \mathbb{P}_a(x_t \in E, T > t) + \int_0^t \mathbb{P}_0(x_{t-s} \in E) \mu_a(ds),$$

where $\mu_a(ds) = \mathbb{P}_a(T \in ds)$ and (using $-E$ instead of E)

$$\mathbb{P}_a(x_t \in -E) = \mathbb{P}_a(x_t \in -E, T > t) + \int_0^t \mathbb{P}_0(x_{t-s} \in -E) \mu_a(ds).$$

Now, $E \subset (0, \infty)$ so x_t cannot belong to $-E$ if $T > t$, i.e., $\mathbb{P}_a[x_t \in -E, T > t] = 0$. Also, $\mathbb{P}_0(x_{t-s} \in -E) = \mathbb{P}_0(x_{t-s} \in E)$. The last

two equalities thus imply
$$\mathbb{P}_a(x_t \in E, T > t) = \mathbb{P}_a(x_t \in E) - \mathbb{P}_a(x_t \in -E).$$

This is what we set out to show. Taking $E = (0, \infty)$, we get the distribution of T:

$$\mathbb{P}_a(T > t) = \int_0^\infty \frac{1}{\sqrt{2\pi t}} \left(e^{-(x-a)^2/2t} - e^{-(x+a)^2/2t} \right) dx$$

$$= \frac{2}{\sqrt{2\pi t}} \int_0^a e^{-x^2/2t} dx = \mathbb{P}_0[|x_t| < a].$$

In particular, we see that $\mathbb{P}_a[T < \infty] = 1$ and that

$$\mathbb{E}_a(T) = \int_0^\infty \mathbb{P}_a[T > t] dt = 2 \int_0^a dx \int_0^\infty \frac{1}{\sqrt{2\pi t}} e^{-x^2/2t} dt = \infty.$$

As another application, consider the d-dimensional Brownian motion. Equation (3.2.8) is clearly equivalent to

$$\mathbb{E}_a[f(x_t)] = \mathbb{E}_a[f(x_t) : T > t] + \int_{[0,t] \times \mathbb{R}^d} \mathbb{E}_b[f(x_{t-s})] \mu_a(dsdb).$$

Let $f(b) = \sum_1^d b_i^2$. Take $a = 0$. We have

$$\mathbb{E}_b[f(x_s)] = \|b\|^2 + d \cdot s.$$

Let T be the exit time through a sphere of radius r:
$$T = \inf\{t : \|x_t\| > r\}$$

and $T = \infty$ if no such t exists. We get

$$d \cdot t = \mathbb{E}_0[f(x_t) : T > t] + \int_{[0,t] \times \mathbb{R}^d} \left[\|b\|^2 + d(t-s) \right] \mu_a(dsdb).$$

Since x_T is clearly on the surface of the sphere $\|b\|^2 = r$, we get

$$d \cdot t = \mathbb{E}_0[f(x_t) : T > t] + r^2 + d\mathbb{E}_0[t - T : T \le t],$$

i.e.,

$$d \cdot t\mathbb{P}_0[T > t] + d\mathbb{E}_0[T : T \le t] = \mathbb{E}_0\left[\|x_t\|^2 : T > t\right] + r^2 \le r^2 + r^2 = 2r^2$$

since $T > t$, $\|x_t\|^2 < r^2$. Letting $t \to \infty$, we see that $t\mathbb{P}_0[T > t]$ is bounded, i.e., $\mathbb{P}_0[T < \infty] = 1$. Then letting $t \to \infty$, we see that $\mathbb{E}_0[T] < \infty$. Finally,

$t \to \infty$ gives $\mathbb{E}_0[T] = \frac{r^2}{d}$. Thus, the first exit time through a sphere of radius r has expectation $\frac{r^2}{d}$.

We note a fundamental property of the d-dimensional Brownian motion:

Fact. *If T is the exit time from a sphere of radius r, center zero, then x_T is uniformly distributed on the surface of the sphere.*

To prove this, note that for any rotation O and $a \in \mathbb{R}^d$,

$$\mathbb{P}_a\left[Ox_{t_i} \in E_i, 1 \le i \le n\right] = \mathbb{P}_{O^{-1}a}\left[x_{t_i} \in E_i, 1 \le i \le n\right]$$

showing that x_t and Ox_t have the same finite-dimensional distributions relative to $\mathbb{P}_0$. Therefore, x_t and Ox_t have the same distribution and this means that x_T is uniformly distributed on the surface of the sphere.

Exercises

3.2.1. Show that θ_1 and α_1 are measurable.

3.2.2. Show that α_t generates the Borel field $\mathscr{B}_t$: The smallest field relative to which α_t is measurable is $\mathscr{B}_t$.

3.2.3. Show that the Brownian motion has no memory (the past and future are independent given the present):

$$\mathbb{P}_a\left[(\theta_t^{-1}A)(\alpha_t^{-1}B) \mid x_t\right] = \mathbb{P}_a\left[(\theta_t^{-1}A) \mid x_t\right]\mathbb{P}_a\left[(\alpha_t^{-1}B) \mid x_t\right]$$

for all $A, B \in \mathscr{B}_\infty$.

Hint: If F and G are bounded $\mathscr{B}_\infty$-measurable functions,

$$\begin{aligned}
\mathbb{E}_a[F \circ \theta_t \cdot G \circ \alpha_t \mid x_t] &= \mathbb{E}_a[G \circ \alpha_t \mathbb{E}_a[F \circ \theta_t \mid \mathscr{B}_t] \mid x_t]\\
&= \mathbb{E}_a[G \circ \alpha_t \mathbb{E}_{x_t}(F) \mid x_t]\\
&= \mathbb{E}_{x_t}(F)\mathbb{E}_a[G \circ \alpha_t \mid x_t]\\
&= \mathbb{E}_a[F \circ \theta_t \mid x_t] \cdot \mathbb{E}_a[G \circ \alpha_t \mid x_t].
\end{aligned}$$

3.2.4. To prove Theorem 3.2.1, it is sufficient to show that

$$\mathbb{E}_a\left[f(x_{T+t}); A \cap (T < \infty)\right] = \mathbb{E}_a\left[\mathbb{E}_{x_t}[f(x_t)]; A \cap (T < \infty)\right],$$

where $a \in \mathbb{R}^1$, $A \in \mathscr{B}_{T+}$, $0 \le f \le 1$, $f \in C(\mathbb{R}^1)$, and $t \ge 0$.

Hint: Use induction as in the case of simple Markov property.

3.2.5. $(T_1 < T_2) \in \mathscr{B}_{T+}$.

Hint: $(T_1 < T_2) \cap (T_1 < t) = \bigcup_{s<t}((T_1 = s) \cap (s < T_2))$.

3.2.6. Show that for a strict Markov time T, $\mathscr{B}_T$ and $\mathscr{B}_{T+}$ are equivalent where

$$\mathscr{B}_{T+} = \bigcap_n \mathscr{B}_{T+\frac{1}{n}} = \{E : E \cap (T \leq t) \in \mathscr{B}_{t+}, \, t \geq 0\}.$$

Hint: Use Proposition 3.2.7.

3.2.7. Show that for any bounded $\mathscr{B}_\infty$-measurable function f,

$$\mathbb{E}_a \left[f\left(\theta_s w\right) \mid \mathscr{B}_{t+s} \right] = Y\left(\theta_s w\right),$$

where

$$Y(w) = \mathbb{E}_a \left[f \mid \mathscr{B}_t \right](w).$$

Hint: First, let f have the form

$$f_1\left(x_{t_1}\right) \cdots f_n\left(x_{t_n}\right) g_{n+1}\left(x_{t_{n+1}}\right) \cdots g_m\left(x_{t_m}\right)$$

with $t_1 < \cdots < t_n < t+s < t_{n+1} < \cdots < t_m$.

3.2.8. Let S be a strict Markov time and T a Markov time such that $T \geq S$. There exists a $\mathscr{B}_S \times \mathscr{B}_\infty$-measurable function $\overline{T}(w_1, w_2)$ on $W \times W$ such that

1. $T(w) = S(w) + \overline{T}(w, \theta_S s)$,
2. $\overline{T}(w, \cdot)$ is a Markov time for each fixed w.

Solution. According to Proposition 3.2.7, the map

$$w \longmapsto (\alpha_S w, \theta_S w) \in W \times W, \quad \mathscr{B}_S \times \mathscr{B}_\infty$$

defined on the set $(S < \infty)$ generates the Borel field $\mathscr{B}_\infty$ restricted to $(S < \infty)$. Thus, there exists a $\mathscr{B}_S \times \mathscr{B}_\infty$-measurable function $\overline{T}(w_1, w_2)$ such that

$$\overline{T}(w) - S(w) = \overline{T}(\alpha_S w, \theta_S w) \quad \text{if } S(w) < \infty.$$

By redefining $\overline{T}(w_1, w_2) = \infty$ if $S(w) = \infty$ or if $w_2(0) \neq w_1(0)$, we do not change (a) or the $\mathscr{B}_S \times \mathscr{B}_\infty$-measurability. This we do. Since $\overline{T}$ is

$\mathscr{B}_S \times \mathscr{B}_\infty$-measurable,
$$\overline{T}(w_1, w_2) = \overline{T}(\alpha_S w_1, w_2).$$

Fix w_1. We must show that
$$(w_2 = \overline{T}(w_1, w_2) = t) \in \mathscr{B}_u \quad \text{for all } u > t,$$
i.e.,
$$\overline{T}(w_1, w_2) = t \quad \text{iff} \quad \overline{T}(w_1, \alpha_u w_2) = t.$$

Define w and w' by
$$w(v) = w_1(v), \quad v \leq S,$$
$$w(v + S) = w_2(v), \quad v \geq 0,$$
$$w' = \alpha_{u+S(w_1)}(w).$$

Clearly, $S(w) = S(w_1) = S(w')$. We have
$$t = \overline{T}(w_1, w_2) = \overline{T}(\alpha_S w_1, w_2) = \overline{T}(\alpha_S w_1, \theta_S w_2) = T(w) - S(w).$$

Thus,
$$T(w) = S(w) + t < S(w) + u.$$

Hence,
$$T(w') = T(w) = S(w) + t = S(w') + t,$$
i.e.,
$$\overline{T}(\alpha_S w', \theta_S w') = t,$$
i.e.,
$$\overline{T}(\alpha_S w_1, \alpha_u w_2) = t$$
since $\alpha_S w' = \alpha_S w_1$, $\theta_S w' = \alpha_u w_2$.

3.2.9. Show that for any t, $0 \leq t < \infty$, the conditional probability of $\mathscr{B}_\infty$ given by $\mathscr{B}_t$ exists, i.e., there exists a function $P(w, B)$ such that we have the following:

1. $P(w, B)$ is $\mathscr{B}_t$-measurable for all $B \in \mathscr{B}_\infty$.
2. $P(w, B)$ is a probability measure for all $w \in W$.
3. $P(w, B) = 1_B(w)$ if $B \in \mathscr{B}_t$.
4. For any a, $B \in \mathscr{B}_t$, $A \in \mathscr{B}_\infty$

$$\mathbb{E}_a\left[P(w, A) : B\right] = \mathbb{P}_a\left[B \cap A\right].$$

Solution. Consider the map

$$w \longmapsto (\alpha_t w, \theta_t w) \in W \times W. \qquad (3.2.9)$$

By Proposition 3.2.7, the least Borel algebra relative to which this map is measurable is precisely $\mathscr{B}_\infty$. Let Ω be the subset of $W \times W$ satisfying

$$(w_1, w_2) \in \Omega \quad \text{iff} \quad w_1 = \alpha_t w_1, \quad w_2(0) = w_1(t).$$

Then the map (3.2.9) maps W onto Ω; indeed, if $(w_1, w_2) \in \Omega$, define $w \in \Omega$ by

$$w(s) = w_1(s), \quad s \le t,$$
$$w(t+s) = w_2(s), \quad s \ge 0.$$

Also, it is clear that Ω is a measurable subset of $W \times W$. It follows that given $B \in \mathscr{B}_\infty$, there exists a unique $\widetilde{B}$ (measurable) $\subset \Omega$ such that

$$B = \left\{w : (\alpha_t w, \theta_t w) \in \widetilde{B}\right\}.$$

For each $w \in W$, define

$$B_w = \left\{w' : (\alpha_t w, w') \in \widetilde{B}\right\}.$$

Then, $B_w \in \mathscr{B}_\infty$. Finally, put

$$P(w, B) = \mathbb{P}_{x(t)}(B_w).$$

We note that if $B \in \mathscr{B}_t$, then $B_w = \emptyset$ if $w \notin B$, $B_w = \{w' : w'(0) = w(t)\}$ if $w \in B$. Thus,

$$P(w, B) = \mathbb{P}_{x(t)}(B_w) = \mathbb{P}_{w(t)}\left(\{w' : w'(0) = w(t)\}\right) = 1_B.$$

Thus, (3) is verified. Equations (1) and (2) we leave to the reader. Equation (4) is verified by looking first on sets determined by time points

$s_1 < \cdots < s_n \leq t < t + t_1 < \cdots < t + t_m$ and then generalizing. Thus, let A be the set

$$\{w : x_{s_i} \in A_i, 1 \leq i \leq n, x_{t+t_j} \in C_j, 1 \leq j \leq m\}.$$

Then, $\tilde{A} \subset \Omega$ is the set

$$(w_1, w_2) \quad \text{such that } w_1 = \alpha_t(w_1),$$
$$w_2(0) = w_1(t), \quad w_1(s_i) \in A_i, \quad 1 \leq i \leq n,$$
$$w_2(t_j) \in C_j, \quad 1 \leq j \leq m.$$

It follows that

$$A_w = \begin{cases} \emptyset & w \notin B_1 = (x_{s_i} \in A_i, 1 \leq i \leq n) \\ B_2 & w \in B_1, \end{cases}$$

where

$$B_2 = \{w' : w'(t_j) \in C_j, 1 \leq j \leq m\}.$$

And if $B \in \mathscr{B}_t$,

$$\mathbb{E}_a\left[P(w, A) : B\right] = \mathbb{E}_a\left[\mathbb{P}_{x(t)}(B_2) : B_1 \cap B\right]$$
$$= \mathbb{P}_a\left[\theta_t^{-1}(B_2) \cap B_1 \cap B\right] = \mathbb{P}_a[A \cap B]$$

since $\theta_t^{-1}(B_2) \cap B_1 = A$.

3.2.10. If R, S are Markov times, then $R(\alpha_S) \geq \min(R, S)$ and $R(\alpha_S)$ is a Markov time.

Hint: For $s > t$, $\mathscr{B}_s \supset (T \leq t) = (T(\alpha_s \leq t))$ for any Markov time T, by Proposition 3.2.2. Using this,

$$(R(\alpha_S) \leq t) = \bigcup_{s \leq t} (R(\alpha_S) \leq t, S = s) \bigcup (R \leq t, S > t),$$

which gives both the assertions.

3.2.11. For any set A, let

$$T = \inf\{t : t \geq 0, x_t \in A\}$$

and $T = \infty$ if no such t exists.

Show that T is a strict Markov time if A is closed and *not* a strict Markov time if A is open.

Hint: Use (3.2.5).

3.2.12. Let D be a bounded open subset of $\mathbb{R}^d$ and T the exit time from D:

$$T = \inf\{t : t > 0, x_t \notin D\},$$

the infimum over an empty set being always ∞ by definition.

Show that

$$\sup_{a \in D} \mathbb{E}_a\left[e^{\epsilon T}\right] < \infty \quad \text{for some } \epsilon > 0.$$

Hint: Let $\varphi(t) = \sup_{a \in D} \mathbb{P}_a[T > t]$. Using Markov property, show that $\varphi(t+s) \leq \varphi(t)\varphi(s)$. So if $\varphi(t_0) < 1$ for one t_0, $\varphi(t) \leq Ke^{-t\lambda}$ for some K, $\lambda > 0$. By considering a ball containing D, it is seen that $\sup_{a \in D} \mathbb{E}_a[T] < \infty$, which implies $\varphi(t_0) < 1$ for large t_0.

For the general theory discussed above, we give the following citations [18–20].

References

[1] Xiaohui Ai and Yang Sun. Karhunen-Loeve expansion for the additive two-sided Brownian motion. *Commun. Stat. Theory Methods*, 47(13): 3085–3091, 2018.

[2] D. Alpay and P. Jorgensen. Transfer operators and conditional expectations: The non-commutative case, the case of mu-Brownian motions and white noise space setting. *Banach J. Math. Anal.*, 18(1): 5, 2024.

[3] R. Gross. Brownian motion can feel the shape of a drum. *Stochastic Process. Appl.*, 167: 104233, 2024.

[4] K. Itô and H. P. McKean, Jr. *Diffusion Processes and Their Sample Paths*. Die Grundlehren der mathematischen Wissenschaften, Band 125. Academic Press, Inc., Publishers, New York; Springer-Verlag, Berlin, 1965.

[5] Frank B. Knight. *Essentials of Brownian Motion and Diffusion*. Mathematical Surveys, No. 18. American Mathematical Society, Providence, 1981.

[6] F. B. Knight. On Brownian motion and certain heat equations. *Z. Wahrsch. Verw. Gebiete*, 55(1): 1–10, 1981.

[7] F. B. Knight. On the upcrossing chains of stopped Brownian motion. In *Séminaire de Probabilités, XXXII*. Lecture Notes in Mathematics, Vol. 1686, pp. 343–375. Springer, Berlin, 1998.

[8] J.-F. Le Gall. *Measure Theory, Probability, and Stochastic Processes*. Graduate Texts in Mathematics, Vol. 295. Springer, Cham, 2022.

[9] A. F. Macedo-Junior and A. M. S. Macêdo. Brownian-motion ensembles of random matrix theory: A classification scheme and an integral transform method. *Nuclear Phys. B*, 752(3): 439–475, 2006.

[10] H. P. McKean, Jr. *Stochastic Integrals*. Probability and Mathematical Statistics, No. 5. Academic Press, New York, 1969.

[11] V. Mitic, G. Lazovic, D. Milosevic, C.-A. Lu, J. Manojlovic, S.-C. Tsay, S. Kruchinin, and B. Vlahovic. Brownian motion and fractal nature. *Modern Phys. Lett. B*, 34(19–20): 2040061, 11, 2020.

[12] E. Nelson. *Dynamical Theories of Brownian Motion*. Princeton University Press, Princeton, 1967.

[13] E. Nelson. Construction of quantum fields from Markoff fields. *J. Funct. Anal.*, 12: 97–112, 1973.

[14] E. Nelson. Connection between Brownian motion and quantum mechanics. In *Einstein Symposion, Berlin (1979)*. Lecture Notes in Physics, Vol. 100, pp. 168–179. Springer, Berlin, 1979.

[15] E. Nelson. Stochastic mechanics and random fields. In *École d'Été de Probabilités de Saint-Flour XV–XVII, 1985–87*. Lecture Notes in Mathematics, Vol. 1362, pp. 427–450. Springer, Berlin, 1988.

[16] E. Nelson. Stochastic mechanics of particles and fields. In *Quantum Interaction*. Lecture Notes in Computer Science, Vol. 8369, pp. 1–5. Springer, Heidelberg, 2014.

[17] M. Radice. First-passage functionals of Brownian motion in logarithmic potentials and heterogeneous diffusion. *Phys. Rev. E*, 108(4): 044151, 2023.

[18] L. Breiman. *Probability*. Addison-Wesley Publishing Co., Reading, 1968.

[19] J. L. Doob. *Stochastic Processes*. John Wiley & Sons, Inc., New York, Chapman & Hall, Ltd., London, 1953.

[20] J. Lamperti. *Probability. A Survey of the Mathematical Theory*. W. A. Benjamin, Inc., New York, 1966.

Chapter 4

Semi-Groups of Operators, Potentials, and Diffusion Equations

With Brownian motion, and an application of Ito's lemma, one can generate solutions to diffusion problems. The purpose of this chapter is to give a presentation of this in the language of semi-groups of bounded operators, parallel with the Hille–Yosida–Phillips theory. This refers to the characterization of the associated infinitesimal generators of strongly continuous one-parameter semi-groups of linear operators acting in Banach spaces.

The main result which we need in the following is often referred to as the Hille–Yosida theorem, and readers are referred to the cited books. See the following. While the full details may be a bit technical, here we stress that the particular semi-groups of operators S_t which we need for this purpose serve as key tools in (i) our analysis of Markov operations from probability, (ii) Ito's lemma from stochastic analysis, and (iii) a more general framework for diffusion equations. The gist of the Hille–Yosida theorem deals with the notion of infinitesimal generator. Every strongly continuous semi-group S_t has a well-defined infinitesimal generator A with dense domain. Moreover, certain *a priori* estimates on the resolvent function computed from A then in fact characterize the operators which are infinitesimal generators. And it will follow that the semi-group S_t may be recovered from A via the corresponding analytic resolvent operator function and an approximation.

The original material in Chapter 3 dealing with semi-groups of linear operators is still current, and we have added a discussion of new directions. In addition, for the benefit of readers, we offer the following citations covering new directions [1, 2, 5, 8–13, 15].

The new directions include fractional semi-groups of operators, reflection positive random fields in Dirichlet spaces, Volterra integrodifferential equations, and reflection positive stochastic processes.

For the benefit of beginner readers, we offer the following supplementary introductory text [3, 4, 7, 8, 14].

Introduction. In this chapter, we give a brief and elementary exposition of semi-group theory. Whereas it is not absolutely essential to know semi-group theory, to read the rest of this book, a knowledge of this theory is indispensable if one wishes to proceed to more general Markov processes.

We give some examples which throw some light on the unifying aspect of the subject. And the patient reader may get an idea of the meaning of boundary conditions in partial differential equations.

Just to start off, the following rough argument tells us how semi-groups naturally enter into considerations of partial differential equations. Suppose we know that for each f in a Banach space of continuous functions on $\mathbb{R}^d$ the equation

$$\frac{\partial u}{\partial t} = Au, \quad u(0, x) = f(x),$$

where A is a differential operator depending only on x has a unique solution. If we put $T_t f(x) = u(t, x)$, then T_t is necessarily a semi-group. Indeed, if $v(t, x) = T_{t+s} f(x) = u(t + s, x)$, then $\frac{\partial v}{\partial t} = Av$ and $v(0, x) = T_s f(x)$. We must thus have, by uniqueness, $T_{t+s} f(x) = T_t [T_s f](x)$.

As said the above is just a rough argument and is not intended to give any idea that miracles will follow. However the semi-group property of the Brownian kernel can already be used to give a simple proof of the

Theorem (Riesz Composition Formula). *If $0 < \alpha, \beta, \alpha + \beta < d$, then*

$$I_\alpha * I_\beta = I_{\alpha+\beta},$$

where for $0 < \theta < d$ the Riesz kernel I_θ is defined by

$$I_\theta(x) = \pi^{-\frac{d}{2}} \frac{\Gamma\left(\frac{d-\theta}{2}\right)}{\Gamma\left(\frac{\theta}{2}\right)} |x|^{-d+\theta}$$

with $|x|$ denoting the norm of $x \in \mathbb{R}^d$.

It is simple to show that $I_\alpha * I_\beta = I_{\alpha+\beta} \cdot (\text{constant})$. The non-trivial part lies in evaluating the constant. M. Riesz evaluated it first and a little later J. Deny used Fourier transforms to give a slightly simpler proof.

Proof of the Riesz Composition Formula. Writing $p(t,x) = (2\pi t)^{-\frac{d}{2}} \exp\left(-\frac{1}{2t}|x|^2\right)$ we have for $0 < \alpha < d$

$$\int_0^\infty t^{\frac{\alpha}{2}-1} p(t,x)\, dt = c_\alpha |x|^{-d+\alpha},$$

where $c_\alpha = 2^{-\frac{\alpha}{2}} \pi^{-\frac{d}{2}} \Gamma\left(\frac{d-\alpha}{2}\right)$. Using the semi-group property of $p(t,\cdot)$, we have thus

$$c_\alpha |x|^{-d+\alpha} * c_\beta |x|^{-d+\beta} = \int_0^\infty t^{\frac{\alpha}{2}-1} dt \int_0^\infty s^{\frac{\beta}{2}-1} p(s+t,x)\, ds$$

$$= \int_0^\infty p(s,x)\, ds \int_0^\infty t^{\frac{\alpha}{2}-1}(s-t)^{\frac{\beta}{2}-1} dt$$

$$= \frac{\Gamma\left(\frac{\alpha}{2}\right)\Gamma\left(\frac{\beta}{2}\right)}{\Gamma\left(\frac{\alpha+\beta}{2}\right)} \int_0^\infty s^{\frac{\alpha+\beta}{2}-1} p(s,x)\, ds.$$

A rearrangement of this equality gives us the composition formula. □

4.1 Semi-Groups

As a very special case of the Markov property, we have

$$\mathbb{E}_a\left[f(X(t+s))\right] = \mathbb{E}_a\left[\mathbb{E}_{X_t}(f(X_s))\right] \tag{4.1.1}$$

for all bounded Borel measurable functions f on $\mathbb{R}^d$. Let T_t denote the operator

$$T_t f(a) = \mathbb{E}_a[f(X_t)], \quad t \geq 0.$$

Equation (4.1.1) can then be written as

$$T_{t+s} = T_t T_s,$$

i.e., the operators T_t form a semi-group. Written in terms of measures, the above semi-group property is a special case of the so-called Chapman–Kolmogorov equation.

Let B be a Banach space. If $u(t)$, $a \leq t \leq b$, is a map on $[a,b]$ into B, we denote by $\int_a^b u(t)\, dt$ the limit when it exists of $\sum_0^n u(t_k)(t_k - t_{k-1})$, where $a = t_0 < t_1 < \cdots < t_m = b$ and as $\max|t_k - t_{k-1}|$ tends to zero; limit here means strong limit, i.e., limit in norm. This definition parallels the definition of the Riemann integral of a function on $[a, b]$. The integral over an infinite interval is then defined in the usual way, i.e., the integral over an arbitrary compact interval in the complement of $[a, b]$ is small provided a and b are chosen suitably. Let us simply write down the following three properties of integral we need. These are all simple to verify:

1. If $u(t)$ is continuous in a compact interval $[a, b]$, then $\int_a^b u(t)\, dt$ exists.
2. $\left\| \int_a^b u(t)\, dt \right\| \leq \int_a^b \|u(t)\|\, dt$ ($\|x\|$ is the norm of the element $x \in B$).
3. If L is in B^*, i.e., L is continuous linear on B, then $L\left(\int_a^b u(t)\, dt\right) = \int_a^b L(u(t))\, dt$.

Let T_t, $t \geq 0$, be a semi-group of linear operators on B. We assume that T_t is *strongly continuous*:

$$\lim_{t \to t_0} \|T_t x - T_{t_0} x\| = 0.$$

And $T_0 = I =$ identity, for all $x \in B$ and $t_0 \geq 0$. It can be shown that (Exercise 4.1.8) a strongly continuous semi-group T_t satisfies $\|T_t\| \leq Me^{\beta t}$ for some $M, \beta > 0$. By considering $e^{-\beta t} T_t$ in place of T_t, we may and do assume from now that $\sup_t \|T_t\| < \infty$.

Example 4.1.1. Using the Brownian motion semi-group, it is simple to derive several examples of semi-groups. For example, let X_t be the one-dimensional Brownian motion. If f is uniformly continuous and bounded on $[0, \infty]$, define

$$S_t f(a) = \mathbb{E}_a [f(|X_t|)].$$

S_t is then a semi-group on the Banach space of such functions and corresponds to the so-called reflected Brownian motion. As another example, let f be continuous on the interval $[0, a]$ with $f(0) = f(a)$. Extend f to $(-a, a)$ to be even: $f(-x) = f(x)$, and now extend it periodically (with period $2a$) to all of $\mathbb{R}$ and then define

$$H_t f(a) = \mathbb{E}_a [f(X_t)].$$

H_t is a semi-group on the Banach space of all those continuous functions f on $[0, a]$ such that $f(0) = f(a)$. This is the semi-group corresponding to the Brownian motion "reflected at 0, and at a".

If f is continuous on $(0, \infty)$ with $f(0) = 0$, we can extend f to an odd function by defining $f(-x) = -f(x)$. Define A_t on such functions by

$$A_t f(a) = \mathbb{E}_a [F(X_t)],$$

where F is the odd extension of f. A_t is then a semi-group on the Banach space of continuous functions f on $[0,\infty]$ such that $f(0) = 0$. This is the semi-group that corresponds to "absorbing barrier at 0". Similarly, if f is a function on $(0,a)$ such that $f(0) = f(a) = 0$, we can extend f to $(-a,a)$ by setting $f(-x) = -f(x)$. Then extend f periodically with period $2a$. And define

$$B_t f(b) = \mathbb{E}_b[F(X_t)], \quad 0 < b < a,$$

where F is the periodic function with period $2a$ obtained as above. B_t is the semi-group that corresponds to absorbing barriers at 0 and a.

Example 4.1.2. Let H be a separable Hilbert space $\{e_n : n \in \mathbb{N}\}$ a complete orthonormal system where $\mathbb{N}$ denotes either the set of all integers or the set of non-negative integers. If λ_n, $n \in \mathbb{N}$, are complex numbers with non-negative real parts, define

$$T_t x = \sum_{n \in \mathbb{N}} e^{-\lambda_n t} \langle x, e_n \rangle e_n,$$

where $\langle x, y \rangle$ denotes the scalar product in H. As special cases of this, one may take the trigonometric system in $(-\pi, \pi)$, the Hermite polynomials in $(-\infty, \infty)$ (with the weight function $e^{-x^2/4}$), the Legendre polynomials in $(-1,1)$, etc.

Example 4.1.3. Let H be a separable Hilbert space and A a self-adjoint operator (bounded or unbounded) with domain $D(A) \subset H$. If E_λ, $-\infty < \lambda < \infty$, is the resolution of identity determined by A, the operators T_t, $t \geq 0$, defined by

$$T_t = \int e^{it\lambda} dE_\lambda$$

form a semi-group on H. (If A is bounded, $T_t = e^{tA}$.)

Example 4.1.4 (Kac's Semi-Group). Let X_t be the d-dimensional Brownian motion, $K > 0$ a bounded measurable function on $\mathbb{R}^d$. Let

$$B = \{f : f \text{ uniformly continuous bounded on } \mathbb{R}^d\}.$$

B provided with the uniform norm is a Banach space. Define for $t \geq 0$

$$T_t f(a) = \mathbb{E}_a\left[f(X_t)\exp\left(-\int_0^t K(X_s)\,ds\right)\right].$$

T_t is a semi-group on B. It is a little tricky to show that T_t maps B into B. See Exercise 4.2.8.

Exercises

4.1.1. Show that Example 4.1.1 are semi-groups and find their "transition densities".

Hint: If $T_t f(x) = \int f(y) p(t, x-y) dy$, where $p(t, z) = \frac{1}{\sqrt{2\pi t}} \exp(-z^2/2t)$, show that f even (odd) implies $T_t f$ is even (odd) and that if f has period a so does $T_t f$ for all $t \geq 0$. With notations in Example 4.1.1, we have
$$S_t f(x) = \int_0^\infty f(y) q(t, x, y) dy, \text{ where}$$
$$q(t, x, y) = p(t, x-y) + p(t, x+y), \quad x > 0,$$
$$H_t f(x) = \int_0^a f(y) h(t, x, y) dy, \text{ where}$$
$$h(t, x, y) = \sum_{-\infty}^{\infty} (p(t, x-y-2ka) + p(t, x+y-2ka)),$$
$$A_t f(x) = \int_0^\infty f(y) a(t, x, y) dy, \text{ where}$$
$$a(t, x, y) = p(t, x-y) - p(t, x+y),$$
$$B_t f(x) = \int_0^a f(y) b(t, x, y) dy, \text{ where}$$
$$b(t, x, y) = \sum_{-\infty}^{\infty} (p(t, x-y-2ka) - p(t, x+y-2ka)).$$

4.1.2. Let B be the Banach space of continuous functions on $[0, 1]$ and let for $f \in B$ and $t > 0$
$$T_t f(x) = \frac{1}{\Gamma(t)} \int_0^x (x-y)^{t-1} f(y) dy, \quad t > 0.$$
Show that $T_{t+s} = T_t T_s$, $\|T_t f - T_s f\| \to 0$ as $s \to t > 0$ but that $T_t f$ does not tend to f in B as t tends to 0 unless $f(0) = 0$.

4.1.3. Complete details in the following. Let T_t be the d-dimensional Brownian semi-group, i.e., $T_t f(a) = \mathbb{E}_a [f(X_t)]$, where X_t is the d-dimensional Brownian motion. If f depends only on distance, i.e., $f(x) = f(|x|)$, $T_t f$ also depends only on distance. Therefore, the definition
$$B_t f(a) = \mathbb{E}_a [f(|X_t|)],$$
where a denotes both a and the vector $(a, 0, 0, \ldots, 0)$, defines a strongly continuous semi-group on the set of bounded uniformly continuous functions on $[0, \infty)$.

4.1.4. There is a well-known procedure to construct new semi-groups from given ones, the so-called subordination procedure. Let F_t be a semi-group of probability measures on $[0, \infty)$: $F_{t+s} = F_t * F_s$, $*$ denoting convolution. Assume that $F_t \to \delta_0$ weakly as $t \to 0$. If T_t is a strongly continuous semi-group on a Banach space, the "subordinated semi-group" defined by

$$S_t = \int_0^\infty T_s F_t(ds)$$

is also strongly continuous. As a special case let $p(t, \cdot)$ be the density of the d-dimensional Brownian semi-group and $F_t(ds)$ the semi-group of Γ-distributions:

$$F_t(ds) = 2^{-\frac{t}{2}} \frac{1}{\Gamma\left(\frac{t}{2}\right)} e^{-\frac{1}{2}s} s^{\frac{t}{2}-1} ds,$$

we obtain the densities

$$S_t(a) = \frac{|a|^{\frac{t-d}{2}} \pi^{-\frac{d}{2}} 2^{-\frac{d+t-2}{2}}}{\Gamma\left(\frac{t}{2}\right)} K_{\frac{t-d}{2}}(a),$$

where $K_\nu(x)$ is the modified Hankel function $= \frac{\pi}{2} \frac{I_{-\nu}(x) - I_\nu(x)}{\sin(\nu\pi)}$ and I_ν is the modified Bessel function

$$I_\nu(x) = \sum_0^\infty \frac{1}{n!\, \Gamma(\nu+n+1)} \left(\frac{x}{2}\right)^{\nu+2n}.$$

The densities S_t were first introduced by Aronszajn and Smith under the name of Bessel potentials in connection with differential problems.

Hint: From Tables of Laplace Transforms Roberts and Kaufman Saunders Co. 1966,

$$\int_0^\infty e^{-\alpha t} t^\nu e^{-a/t} dt = 2 \left(\frac{a}{\alpha}\right)^{\frac{\nu+1}{2}} K_{\nu+1}\left(2a^{1/2}\alpha^{1/2}\right)$$

$\Re a > 0$, $\Re \alpha > 0$.

4.1.5. If m is a probability measure on $\mathbb{R}^d$, show that $m * f(x) = \int f(x-y) m(dy)$ exists in L^p, $1 \leq p \leq \infty$, for all Borel measurable $f \in L^p$ and ($\|\cdot\|_p$ denoting norm) $\|m * f\|_p \leq \|f\|_p$.

Hint: Suppose $f \geq 0$. For all $g \in L^q$,

$$\int g(x) dx \int f(x-y) m(dy) = \int m(dy) \int g(x) f(x-y) dx$$

$$\leq \|g\|_q \|f\|_p$$

proving that $\int f(x-y) m(dy) \in L^p$.

4.1.6. Let F_t, $t \geq 0$, be probability measures on $\mathbb{R}^d$ such that $F_t * F_s = F_{t+s}$ and $\lim_{t \to 0} F_t * f = f$, pointwise for all bounded continuous f. $T_t f = F_t * f$ then defines a strongly continuous semi-group on $L^p(\mathbb{R}^d)$ for $1 \leq p \leq \infty$.

Hint: If g is continuous, has compact support, and $\|f - g\|_p$ is small, then $\|T_t(f-g)\|_p$ is small for all t. $T_t g \to g$, pointwise as $t \to 0$ and $\|T_t g\|_p \leq \|g\|_p$. Fatou implies $\lim_{t \to 0} \|T_t g\|_p = \|g\|_p$. No use Exercise 4.1.7.

4.1.7. Let $f_n \in L^p$. Suppose $f_n \to f$ almost everywhere, $f \in L^p$ and $\|f_n\|_p \to \|f\|_p$. Then f_n tends to f in L^p.

Hint: Let A_n be the set where $|f_n| \leq 2|f|$ and B_n the complement of A_n. If $g_n = f_n$ on A_n and zero elsewhere, g_n tends to f in L^p by dominated convergence. Therefore, $\int_{B_n} |f_n|^p \, dx$ tends to zero. Finally,

$$\int |f_n - f|^p \, dx \leq \int |g_n - f|^p \, dx + 2 \int_{B_n} |f_n|^p \, dx.$$

4.1.8. Let T_t, $t \geq 0$, be a strongly continuous semi-group on a Banach space B. Show that

1. $\|T_t\|$ is bounded in every compact interval,
2. $\lim_{t \to \infty} \|T_t\|^{\frac{1}{t}} = \inf_t \|T_t\|^{\frac{1}{t}}$.

Thus, there exists M and β such that $\|T_t\| \leq Me^{\beta t}$, $t \geq 0$.

Hint: (1) Put $q(x) = \sup_{t \in K} T_t x$, where K is compact. Then q is a lower semicontinuous semi-norm. Use Baire category theorem to conclude that $q(x) \leq p\|x\|$ for some p.

(2) If a is larger than the inf, choose t_0 so that

$$a > \|T_{t_0}\|^{\frac{1}{t_0}},$$

and for $nt_0 \leq t \leq (n+1)t_0$,

$$\|T_t\|^{\frac{1}{t}} \leq \|T_{t-nt_0}\|^{\frac{1}{t}} a^{\frac{nt_0}{t}}.$$

Now use the fact from (1) that $\|T_t\|$ is bounded for $0 \leq t \leq 1$.

4.2 Infinitesimal Generators

The semi-group property clearly makes semi-groups of operators hard to come by. One has to replace it by a simpler object. And this is the infinitesimal generator.

Let T_t be a strongly continuous semi-group on a Banach space B. We assume that $\|T_t\| \leq M$ for all t.

Definition 4.2.1. The infinitesimal generator A of T_t is defined to be the operator whose domain $D(A)$ is the set of $x \in B$ such that

$$D(A) = \left\{ x : \lim_{t \to 0} \frac{T_t x - x}{t} \text{ exists} \right\}$$

and this limit is by *definition* Ax.

If $x \in D(A)$, it is easily verified that $T_t x \in D(A)$ for all t and $\frac{d}{dt} T_t x = T_t A x$.

This suggests that in some sense $T_t = \exp(tA)$. If A were bounded, the right side is meaningful. However, this is still true in a limiting sense. See Exercise 4.2.2.

In the investigation of semi-groups, a fundamental role is played by the *Resolvent Operator* which is defined for $\lambda > 0$ by

$$R_\lambda x = \int_0^\infty e^{-\lambda t} T_t x \, dt.$$

From our condition that $\|T_t\| \leq M$ for all t, it follows that $\lambda \|R_\lambda\| \leq M$. An induction argument shows that

$$R_\lambda^{n+1} x = \frac{1}{n!} \int_0^\infty t^n e^{-\lambda t} T_t x \, dt, \quad n \geq 0,$$

so that $\lambda^{n+1} \|R_\lambda^{n+1}\| \leq M$ for $n \geq 0$. It is very simple to show that R_λ satisfies the resolvent equation:

$$R_\lambda - R_\mu + (\lambda - \mu) R_\lambda R_\mu = 0.$$

The resolvent equation shows that the range of R_λ is independent of λ. Also $R_\lambda x = 0$ for some λ implies, using the resolvent equation, that $R_\mu x = 0$ for all μ, and since $\mu R_\mu x$ tends to x as μ tends to infinity, $x = 0$.

The relation between the infinitesimal generator and the resolvent operators is contained in the following theorem.

Theorem 4.2.2. $D(A) = R_\lambda(B)$ and $A(R_\lambda x) = \lambda R_\lambda x - x$.

Proof. Let $u = R_\lambda x$. Let us show that $u \in D(A)$. We have

$$T_t u - u = \int_0^\infty e^{-\lambda s} T_{t+s} x \, ds - \int_0^\infty e^{-\lambda s} T_s x \, ds$$

$$= (e^{\lambda t} - 1) \int_t^\infty e^{-\lambda s} T_s x \, ds - \int_0^t e^{-\lambda s} T_s x \, ds.$$

Thus,

$$\lim_{t \to 0} \frac{T_t u - u}{t} = \lambda u - x.$$

Conversely, let $u \in D(A)$ and $x = \lim_{t \to 0} \frac{T_t u - u}{t}$. Since R_λ is a bounded operator and R_λ commutes with T_t, we get

$$R_\lambda x = \lim_{t \to 0} \frac{T_t R_\lambda u - R_\lambda u}{t} = \lambda R_\lambda u - u$$

from what we just proved, namely that $R_\lambda u \in D(A)$ and $AR_\lambda u = \lambda R_\lambda u - u$. This shows that $u = R_\lambda(\lambda u - Au)$. □

It is useful to note the following facts which are completely contained in the proof:

1. $\lambda u = Au$ for some λ implies $u = 0$. In other words, $\lambda I - A$ defined on $D(A)$ is 1-1, where I = identity.
2. $\lambda I - A$ is onto. Indeed, for any $x \in B$, $u = R_\lambda x \in D(A)$ and $\lambda u - Au = x$.
3. $u = R_\lambda(\lambda u - Au)$, i.e., $(\lambda I - A)^{-1}$ exists and equals R_λ and we know that $\|R_\lambda^n\| \leq \lambda^{-n} M$.
4. Two semi-groups with the same infinitesimal generator are identical. Indeed, from (3), the semi-groups must have the same resolvent and therefore the uniqueness of Laplace transforms implies the identity of the semi-groups.
5. A is a closed operator: $D(A) \ni u_n \to u$, $Au_n \to y$ imply $u \in D(A)$ and $Au = y$. Indeed, if $u_n = R_1 x_n$, $Au_n = u_n - x_n$. Our assumptions say that u_n and x_n converge. Now recall that R_1 is bounded. In particular, we see from the closed theorem that $D(A) = B$ if and only if A is bounded.

Theorem 4.2.3. *Let B be a Banach space and A a linear operator defined on a dense subspace of B. A is the infinitesimal generator of a semi-group T_t, $\|T_t\| \leq M$ if and only if for all $\lambda > 0$, $(\lambda I - A)$ maps $D(A)$ onto B, $R_\lambda = (\lambda I - A)^{-1}$ exists, and $\|R_\lambda^n\| \leq \lambda^{-n} M$.*

Proof. See Exercise 4.2.6. □

It is not always easy nor essential to know the precise domain of a generator. It is sufficient most often to know "enough" elements of the domain. The following examples illustrate this.

Example 4.2.4. In Example 4.1.1, the semi-groups were constructed using the Brownian semi-group. It is natural that their generators should be expressible in terms of the generator of the Brownian semi-groups plus some conditions at the boundary points. It is very simple to show (using Taylor's expansion) that every C^2-function u with compact support belongs to the domain of generator of the Brownian semi-group and $Au = \frac{1}{2}\frac{d^2u}{dx^2}$.

Retain the notation of Example 4.1.1 f belongs to the domain of generator of S_t if and only if $f(|\cdot|)$ (defined on all $\mathbb{R}^1$) belongs to the domain of generator of the Brownian semi-group. In particular, if f has compact support and $f(|\cdot|)$ is C^2, it will belong to the domain of generator of S_t. Thus, the generator can loosely be described by "$\frac{1}{e}u'''$" with the boundary condition $u'(0) = 0$.

Similar reasoning applies to the other examples.

Example 4.2.5. If T_t and S_t are commuting strongly continuous semi-groups, i.e., $T_tS_s = S_sT_t$ for all $0 \leq x, t$, then $C_t = T_tS_t$ is also a strongly continuous semi-group. The generator C of C_t has domain $D(C) \supset D(T) \cap D(S)$, T, S generators of T_t, S_t, respectively. And for $u \in D(T) \cap D(S)$,

$$Cu = Tu + Su.$$

This simple fact can be used as follows. Let $Z(t) = (x(t), y(y))$ denote the two-dimensional Brownian motion. Let $B =$ the set of bounded uniformly continuous functions on $\mathbb{R}^2$. Let T_t, S_t be the semi-groups on B:

$$T_t f(x, y) = \mathbb{E}_x [f(x_t, y)],$$

$$S_t f(x, y) = \mathbb{E}_y [f(x, y_t)].$$

It is easy to show that T_t and S_t commute. $C_t = T_tS_t$ is simply the two-dimensional Brownian semi-group. Thus, $u \in C^2$ with compact support implies that u is in the domain of generator of C_t and

$$Cu = \frac{1}{2}\left(\frac{\partial^2 u}{\partial x^2} + \frac{\partial^2 u}{\partial y^2}\right).$$

Similarly, if u is C^2 with compact support in $\mathbb{R}^d$, then $u \in D(A)$, $A =$ generator of d-dimensional Brownian motion semi-group, and $Au = \frac{1}{2}\Delta u$, $\Delta =$ Laplacian.

As another example of the same idea, consider
$$\frac{\partial u}{\partial t} = \frac{1}{2}\frac{\partial^2 u}{\partial x^2} - v\frac{\partial u}{\partial x}, \quad u(0,x) = f(x),$$
where v is a constant. This is the equation of diffusion in a rod which moves with velocity v along the x-axis. The semi-group $S_t f(a) = f(a-vt)$ has infinitesimal generator $-v\frac{d}{dx}$ in the sense that if u is differentiable and belongs to the domain of generator A of S_t, then $Au = -v\frac{du}{dx}$. S_t commutes with the one-dimensional Brownian semi-group T_t whose generator is $\frac{1}{2}\frac{d^2}{dx^2}$. Thus,

$$u(t,a) = T_t S_t f(a) = \frac{1}{\sqrt{2\pi t}} \int_{-\infty}^{\infty} f(b-vt) \exp\left(-\frac{(b-a)^2}{2t}\right) db$$

$$= \frac{1}{\sqrt{2\pi t}} \int_{-\infty}^{\infty} f(b) \exp\left(-\frac{(b+vt-a)^2}{2t}\right) db$$

satisfies the above diffusion equation.

Example 4.2.6. Let A be the generator of Kac's semi-group of Example 4.1.4. Assume $K \in B$ (with the notation of the cited example). Using the computation of Exercise 4.2.8, if $v = G_\alpha f$, then $v = R_\alpha(f - Kv)$. Therefore, $Av = \alpha v - f = \alpha R_\alpha(f - Kv) - (f - Kv) = \frac{1}{2}\Delta v - Kv$, where Δ denotes the generator of the Brownian semi-group.

Example 4.2.7. This example illustrates how semi-groups help in solving perturbed equations. Let T_t be a strongly continuous semi-group on a Banach space B. Let A denote the generator of T_t. Suppose $f : [0,\infty) \to B$ is a strongly differentiable map. Using routine computation, for each $x_0 \in D(A)$, the function

$$y(t) = T_t x_0 + \int_0^t T_{t-s} f(s) \, ds$$

satisfies $\frac{dy}{dt}(t) = Ay(t) + f(t)$. Apply this to the d-dimensional Brownian motion semi-group: The solution to

$$\frac{\partial u}{\partial t} = \frac{1}{2}\Delta u + g(t,a), \quad \lim_{t \to 0} u(t,a) - f(a)$$

is given by (for nice $g(t,a)$)

$$u(t,a) = \mathbb{E}_a(f(X_t)) + \int_0^t \mathbb{E}_a(g(s, X_{t-s})) \, ds,$$

where X_t denotes the d-dimensional Brownian motion.

Exercises

4.2.1. Let A be a bounded operator on a Banach space:

$$T_t = \exp(tA) = \sum_0^\infty \frac{t^n}{n!} A^n$$

is then a strongly continuous semi-group with infinitesimal generator A.

4.2.2. Show that

$$Af(x) = \int_0^x (f(y) - f(x)) \, dy,$$

f continuous on $[0,1]$, generates a semi-group T_t. Find T_t.

Hint: A is bounded, and

$$T_t f(x) = e^{-tx} f(x) + t \int_0^x e^{-ty} f(y) \, dy.$$

4.2.3. Let S_t, T_t be commuting strongly continuous semi-groups ($S_s T_t = T_t S_s$ is assumed) with infinitesimal generators A and B. Assume that

$$\|T_t\| \le M, \quad \|S_t\| \le M$$

for all t. Then for $x \in D(A) \cap D(B)$,

$$\|S_t x - T_t x\| \le M^2 t \|Ax - Bx\|.$$

Hint:

$$S_t - T_t = S_{\frac{t}{n}}^n - T_{\frac{t}{n}}^n = \left(S_{\frac{t}{n}} - T_{\frac{t}{n}}\right)\left(S_{\frac{t}{n}}^{n-1} + \cdots + T_{\frac{t}{n}}^{n-1}\right)$$

so that

$$\|S_t x - T_t x\| \le n M^2 \left\| S_{\frac{t}{n}} x - x - \left(T_{\frac{t}{n}} x - x\right) \right\|$$

because

$$\left\|S_{\frac{t}{n}}^i - T_{\frac{t}{n}}^j\right\| = \left\|S_{\frac{it}{n}} - T_{\frac{jt}{n}}\right\| \le M^2.$$

Now use $\frac{S_u x - x}{u}$ tends to Ax as u tends to zero.

4.2.4. Let $\|T_t\| \le M$. Show that $S_{t,h} x = \exp\left(t \frac{T_h - I}{h}\right) x$ converges strongly to $T_t x$ as $h \to 0$.

Hint:

$$\|S_{t,h}\| \le e^{-\frac{t}{h}} \sum \frac{t^n}{h^n n! \|T_h^n\|} \le M.$$

$S_{t,h}$ is a semi-group with generator $\frac{T_h - I}{h}$. From Exercise 4.2.3, $S_{t,h}x$ tends to $T_t x$ (uniformly in compact intervals) for all $x \in D(A)$. And $D(A)$ is dense.

4.2.5. Let R_λ be the resolvent of T_t, where $\|T_t\| \leq M$. Show that $T_{t,\lambda}(x) = \exp(t\lambda(\lambda R_\lambda - I)x)$ tends to $T_t x$ as λ tends to infinity.

Hint: Using $\|(\lambda R_\lambda)^n\| \leq M$, as in Exercise 4.2.4, $\|T_{t,\lambda}\| \leq M$. So $T_{t,\lambda}$ is a semi-group with generator $\lambda(\lambda R_\lambda - I)$. For $x \in D(A)$, $\lambda R_\lambda A x = \lambda A(R_\lambda x) = \lambda(\lambda R_\lambda x - x)$. And $\lambda R_\lambda A x$ tends to Ax as λ tends to infinity. The rest is as in Exercise 4.2.4.

4.2.6. Prove Theorem 4.2.3 in the following steps.

Step 1: If $R_\lambda = (\lambda - A)^{-1}$, R_λ, R_μ commute. Indeed, if $x \in D(A)$, $(\lambda I - A) R_\lambda x = x$ and $(\mu I - A) R_\mu y = y$. Operate on both sides of the first equality by R_μ and let $y = R_\lambda x$ in the second. Subtracting the resulting equations, one obtains $(R_\lambda - R_\mu + (\lambda - \mu) R_\mu R_\lambda) x = 0$. $D(A)$ is dense.

Step 2: Since $\|R_\lambda x\| \leq M \lambda^{-1} \|x\|$, the equality $(\lambda I - A) R_\lambda x = x$ shows that $\lambda R_\lambda x$ tends to x for all x in $D(A)$ and hence for all x.

Step 3: $\lambda R_\lambda A x = \lambda A(R_\lambda x) = \lambda(\lambda R_\lambda - I)x$. So the semi-group $T_{t,\lambda} = \exp(t\lambda R_\lambda A x)$ has norm $\leq M$. $T_{t,\lambda}$ and $T_{s,\mu}$ commute from Step 1. And

$$\|T_{t,\lambda} x - T_{t,\mu} x\| \leq M^2 t \|\lambda R_\lambda A x - \mu R_\mu A x\|.$$

Conclude from Step 2 that $\lim_{\lambda \to \infty} T_{t,\lambda} x = T_t x$ exists uniformly for t in a compact interval. T_t is then a strongly continuous semi-group. Letting μ tends to infinity, we get, for all $x \in D(A)$,

$$\|T_{t,\lambda} x - T_t x\| \leq M^2 t \|\lambda R_\lambda A x - A x\|.$$

Step 4: Let A_1 be the generator of T_t. Use the last inequality in Step 3 to show that $x \in D(A)$ implies $x \in D(A_1)$ and $A_1 x = A x$.

Step 5: Conclude $D(A_1) = D(A)$. Indeed, if $R_\lambda^1 = (\lambda - A_1)^{-1}$, $x = (\lambda - A_1) R_\lambda^1 x$, $x = (\lambda - A_1) R_\lambda x$, since by Step 4 $A_1 = A$ for $y \in D(A)$. By uniqueness, $R_\lambda x = R_\lambda^1 x$.

4.2.7. If f is C^∞ on the real line with compact support in the open interval $(0, \infty)$,

$$u = \int_0^\infty f(t) T_t x \, dt \in D(A)$$

and

$$Au = -\int_0^\infty f'(t) T_t x \, dt.$$

Semi-Groups of Operators, Potentials, and Diffusion Equations 57

In particular, $D_\infty(A)$ is dense in B where $D_1(A) = D(A)$ and for $k \geq 2$, $D_k(A) = \{u : u \in D_{k-1}(A), Au \in D(A)\}$ and $D_\infty(A) = \bigcap_k D_k(A)$.

4.2.8. Show that Kac's semi-group, the semi-group of Example 4.1.4, maps B into B.

Hint: That T_t is a semi-group follows from the Markov property. Write

$$R_\alpha f(a) = \mathbb{E}_a \left[\int_0^\infty e^{-\alpha t} f(x_t) \, dt \right],$$

$$G_\alpha f(a) = \int_0^\infty e^{-\alpha t} \mathbb{E}_a \left[f(x_t) \exp\left(-\int_0^t K(x_s) \, ds\right) \right] dt.$$

Using

$$1 - \exp\left(-\int_0^t K(x_s) \, ds\right) = \int_0^t K(x_s) \exp\left(-\int_s^t K(x_\theta) \, d\theta\right) ds$$

and Fubini,

$$R_\alpha f(a) - G_\alpha f(a)$$

$$= \mathbb{E}_a \left[\int_0^\infty K(x_s) \, ds \int_s^\infty e^{-\alpha t} f(x_t) \exp\left(-\int_s^t K(x_\theta) \, d\theta\right) dt \right]$$

(change of variables and use the Markov property)

$$= \mathbb{E}_a \left[\int_0^\infty K(x_s) e^{-\alpha s} v(x_s) \, ds \right] = R_\alpha [Kv],$$

where $v(a) = G_\alpha f(a)$. Now both $R_\alpha f$ and $R_\alpha[Kv]$ are in B. (If $\delta > 0$, $|\mathbb{E}_a(g(x_t)) - \mathbb{E}_{a+h}(g(x_t))| < \epsilon$ for all $a \in \mathbb{R}^d$ and $t \geq \delta$, provided h is small enough, showing that $R_\alpha g \in B$ for all bounded Borel g.) $G_\alpha f$ is therefore in B. $\alpha G_\alpha f$ converges uniformly to f if $f \in B$. By the Hille–Yosida theorem, Theorem 4.2.2, the resolvent G_α determines a semi-group S_t whose Laplace transform is G_α. Since $T_t f$ is continuous in t for $f \in B$, we must have $T_t f = S_t f$.

4.2.9. Let R_λ be operators on a Banach space B such that $\|\lambda R_\lambda\| \leq 1$. Suppose R_λ satisfies the resolvent equation and the range of R_λ is dense in B. Show that there is a unique contraction semi-group T_t on B whose resolvent is R_λ.

4.2.10. Let B be a Banach space and A an operator with dense domain $D(A)$. A is the infinitesimal generator of a strongly continuous semi-group if and only if there is a $\lambda_0 \geq 0$ such that for $\lambda > \lambda_0$, $R_\lambda = (\lambda I - A)^{-1}$ exists and $\|R_\lambda^n\| \leq M(\lambda - \lambda_0)^{-n}$ for all n.

Hint: Apply the Hille–Yosida theorem to $A - \lambda_0 I$.

4.2.11. Let Δ denote the generator of the Brownian semi-group on $B = C_0(\mathbb{R}^d) =$ the space of continuous functions on $\mathbb{R}^d$ vanishing at ∞. Let p be a bounded continuous function of on $\mathbb{R}^d$. Show that $pI + \Delta$ generates a semi-group on B.

Hint: Let R_λ denote the resolvent of the Brownian kernel. $u \mapsto R_\lambda(pu)$ is an operator on B of norm < 1 for $\lambda > \|p\|$. So $S_\lambda = (I - R_\lambda p)^{-1} R_\lambda$ exists in B for $\lambda > \|p\|$ and $\|S_\lambda^n\| \leq (\lambda - \|p\|)^{-n}$. Finally, $u = S_\lambda f$ solves $\lambda u - pu - \Delta u = f$. Now use Exercise 4.2.10.

4.2.12. Let p_t, $t \geq 0$, be probability measures on $\mathbb{R}^d$ such that $p_t * p_s = p_{t+s}$ and $\lim_{t \to 0} p_t = \delta_0$. Then ψ defined by $e^{-t\psi(\alpha)} = \int \exp(i\langle \alpha, x\rangle) p_t(dx)$ is continuous and satisfies $|\psi(2\alpha)| \leq 4|\psi(\alpha)|$.

In particular, $|\psi(\alpha)| = O(|\alpha|^2)$ as $|\alpha|$ tends to infinity.

Hint: Let $\varphi(\alpha) = \int e^{i\langle \alpha, x\rangle} F(dx)$ be any characteristic function. Use the identity $(1-a) + (1-b) - (1-a)(1-b) = 1 - ab$ and

$$\left|\int \left|1 - e^{i\langle \alpha, x\rangle}\right| F(dx)\right|^2 \leq 2 \int (1 - \cos\langle \alpha, x\rangle) F(dx) \leq 2|1 - \varphi(\alpha)|$$

to show that $p(\alpha) = \sqrt{|1 - \varphi(\alpha)|}$ is sub-additive: $p(\alpha + \beta) \leq p(\alpha) + p(\beta)$. Use this to show that $\sqrt{|\psi(\alpha)|}$ is sub-additive. Continuity of ψ follows from the following: $\frac{1}{1+\psi}$ is the characteristic function of the probability measure $m(dx) = \int_0^\infty e^{-t} p_t(dx) dt$.

4.2.13. Let T_t be a convolution semi-group no $C_0(\mathbb{R}^d)$: $T_t f = p_t * f$, where p_t are as in Exercise 4.2.12. Show that every C^3-function with compact support is in the domain of generator of T_t.

Hint: Retain the notation of Exercise 4.2.12. The domain of generator is precisely the set $\{m * f : f \in C_0(\mathbb{R}^d)\}$. Now use Exercise 12 and Fourier transforms.

For more on convolution semi-groups, see [3].

4.2.14. Let T_t be a positive semi-group on $C_0(\mathbb{R}^d)$ with generator A. Suppose $D(A) \supset \mathscr{D} =$ the set of C^∞-functions with compact support and that support $(Au) \subset $ support (u) for all $u \in \mathscr{D}$. Show that $A|_{\mathscr{D}}$ is a differential operator of order at most 2.

Hint: If $u \in \mathscr{D}$ vanishes together with all its first two partials at a point x_0, then $(Au)(x_0) = 0$. Indeed, if $0 \leq P_\gamma \leq 1$ are in $\mathscr{D}$ such that $P_\gamma = 1$

in say $B(x_0, \gamma)$ = the ball of radius y and center x_0 and with $P_\gamma = 0$ off $B(x_0, 2\gamma)$. Since $u = uP_1P_\gamma$ in $B(x_0, \gamma)$,

$$|Au(x_0)| = |A(uP_1P_\gamma)(x_0)| = \lim_{t \to 0} \left| \frac{T_t(uP_1P_\gamma)(x_0)}{t} \right|$$

$$\leq 2M\gamma \lim_{t \to 0} \frac{T_t\left((x-x_0)^2 P_1(\cdot)\right)(x_0)}{t}$$

since $|u| \leq M|x - x_0|^3$. The last limit is finite because $(x-x_0)^2 P_1 \in \mathscr{D}$.

Now we determine continuous functions a, a_i, a_{ij} as follows: Let $\varphi_n \in \mathscr{D}$ be such that $\varphi_n = 1$ in $B(0,n)$. Then $x_i\varphi_n, x_ix_j\varphi_n$ are in $\mathscr{D}$, where x_i are the coordinate functions. Define

$$a(x) = (A\varphi_n)(x),$$
$$a_i(x) = A(x_i\varphi_n)(x),$$
$$a_{ij}(x) = A(x_ix_j\varphi_n)(x),$$

for $x \in B(0,n)$, $1 \leq i,j \leq d$. Fix x_0 and let $u \in \mathscr{D}$. For all n such that support $(u) \subset B(0,n)$,

$$u(x) = u(x_0)\varphi_n + \sum_i \left(\frac{\partial u}{\partial x_i}(x_0)\right)(x_i - x_{0i})\varphi_n(x)$$

$$+ \frac{1}{2}\sum_{i,j}\left(\frac{\partial^2 u}{\partial x_i \partial x_j}(x_0)\right)(x_i - x_{0i})(x_j - x_{0j})\varphi_n(x)$$

$$+ u(x_0, x)\varphi_n(x),$$

where $u(x_0, \cdot) \in \mathscr{D}$ vanishes together with all its first and second partials at x_0. Therefore, $A[u(x_0, \cdot)\varphi_n](x_0) = 0$ and

$$(Au)(x_0) = a(x_0)u(x_0) + \sum_i b_i(x_0)\frac{\partial u}{\partial x_i}(x_0) + \sum_{i,j} b_{ij}(x_0)\frac{\partial^2 u}{\partial x_i \partial x_j}(x_0),$$

where

$$b_i(x) = a_i(x) - x_{0i}a(x),$$
$$b_{ij}(x) = \frac{1}{2}(a_{ij}(x) - x_{0i}a_j(x) - x_{0j}a_i(x) + x_{0i}x_{0j}a(x)).$$

4.2.15. In contrast to the above exercise, show that Δ^2 is the restriction to $\mathscr{D}$ of the generator of a semi-group on $C_0\left(\mathbb{R}^d\right)$.

Hint: $\exp(-t|x|^4)$ being rapidly decreasing is the Fourier transform of a rapidly decreasing function $F(t,x)$:

$$\int \exp\left(i\left\langle \xi, x\right\rangle F(t,x)\right) dx = \exp\left(-t\left|\xi\right|^4\right).$$

The semi-group

$$T_t f(x) = \int f(y) F(t, x - y) \, dy$$

has as generator a constant multiple of Δ^2.

4.3 Potential Operators

Potential operators are in a general sense inverses of infinitesimal generators. Most important examples of these are Green functions which we discuss in Chapter 7.

Let T_t be a strongly continuous semi-group with $\|T_t\| \leq M$ on a Banach space B. R_λ will be the resolvent of T_t and A its infinitesimal generator. To investigate when A^{-1} exists as a densely defined operator, first consider the equation $Au = 0$. Using the fact that u is in the range of R_λ and the resolvent equation, this is possible only if $u = \lambda R_\lambda u$ for all λ, or equivalently, $T_t u = u$ for all t. Next, suppose $v = Au$ is in the range of A. Then $\lambda R_\lambda v = \lambda R_\lambda A u = \lambda A\left(R_\lambda u\right) = \lambda\left(\lambda R_\lambda u - u\right)$, which clearly tends to zero as λ tends to zero. Since $\|\lambda R_\lambda\| \leq M$, we also have $v \in$ the closure of the range of A if and only if $\lim_{\lambda \to 0} \lambda R_\lambda = 0$. (Indeed, if $\lambda R_\lambda v$ tends to zero as λ tends to zero, $AR_\lambda v = \lambda R_\lambda v - v$ so that $A\left(R_\lambda v\right)$ tends to $-v$ as λ tends to zero.)

Thus, A^{-1} has a densely defined domain if and only if $\lim_{\lambda \to 0} \lambda R_\lambda x = 0$ for all $x \in B$. Further, if $v = Au$,

$$R_\lambda v = R_\lambda A u = A R_\lambda u = \lambda R_\lambda u - u$$

so that $\lim_{\lambda \to 0} R_\lambda v = -u$. Conversely, if $\lim_{\lambda \to 0} R_\lambda v = u$ exists, the resolvent equation $R_1 v - R_\lambda v + (1 - \lambda) R_1 R_\lambda v = 0$ gives $R_1 v - u + R_1 u = 0$, i.e., $u \in D(A)$ and $Au = -v$. We have proved the following.

Proposition 4.3.1. (1) $v \in$ *the closure of the range of A if and only if* $\lim_{\lambda \to 0} \lambda R_\lambda v = 0$. (2) A^{-1} *exists as a densely defined operator if and only if* $\lim_{\lambda \to 0} \lambda R_\lambda x = 0$ *for all* $x \in B$. (3) *If A^{-1} exists as a densely defined* *operator, then $v \in D(A^{-1})$ if and only if $\lim_{\lambda \to 0} R_\lambda v = u$ exists and then $A^{-1}v = -u$.*

Generally speaking, potential operators are integral operators and thus are sometimes easier to handle. For example, consider the d-dimensional Brownian semi-group as acting on the space $C_0(\mathbb{R}^d)$ of continuous functions vanishing at ∞. The above reasoning applies to this and we see that A^{-1} has a densely defined domain. It is shown in Chapter 6 that $u \in C^2$ with compact support then

$$u(x) = A_d \int K(x-y) \Delta u(y) \, dy,$$

where A_d are constants and $K(y) = -\log|y|$ if $d = 2$ and $K(y) = |y|^{-d+2}$ if $d \geq 3$. Thus, at least A^{-1} restricted to $A(\mathscr{D})$ ($\mathscr{D}$ = all C^∞-functions with compact support) is given by an integral operator.

The most important theorem in this connection is a theorem of G. Hunt. To describe this, we need a little terminology. Let $K(X)$ and $C_0(X)$ denote the space of all continuous functions with compact support and all continuous functions vanishing at ∞, respectively, on the locally compact, σ-compact space X. A linear map $V : K(x) \to C_0(X)$ is said to satisfy the *principle of positive maximum* if the following holds:

For every $f \in K(X)$ such that Vf attains strictly positive values,

$$\sup_x Vf(x) = \sup\{Vf(y) : f(y) \geq 0\}. \tag{4.3.1}$$

Equation (4.3.1) is equivalent to the following apparently stronger condition:

For every $f \in K(X)$ such that Vf attains strictly positive values,

$$\sup Vf(x) = \sup\{Vf(y) : f(y) > 0\}. \tag{4.3.2}$$

Indeed, suppose a and b denote the left and right sides of (4.3.2) and $b < c < a$. Put

$$A = \{x : Vf(x) \geq c\}.$$

A is compact since Vf tends to zero at infinity. Let g be any function in $K(X)$, which is strictly negative on A. For ϵ such that $\epsilon \|Vg\| < \frac{a-c}{2}$, for all points $x \notin A$, $V(f+\epsilon g) < \frac{a+c}{2}$, while at any $x \in A$ at which $Vf(x) = a$, $Vf(x) + \epsilon Vg(x) > \frac{a+c}{2}$. This contradicts (4.3.1) since at all points of A, $f + \epsilon g < 0$.

Equation (4.3.2) shows that V must be a non-negative operator. For if $f \leq 0$, the right side of (4.3.4) is zero so that Vf cannot attain a strictly positive value.

Another condition equivalent to (4.3.1) is the following: For all $\alpha \geq 0$, $\lambda > 0$, and $f \in K(X)$,

$$\alpha + \lambda V f + f \geq 0 \quad \text{implies} \quad \alpha + \lambda V f \geq 0. \tag{4.3.3}$$

Indeed, if $\alpha + \lambda V f(x) < 0$, $V(-f)$ would attain strictly positive values and, assuming (4.3.1),

$$\sup_{f \leq 0} \lambda V(-f) = \sup \lambda V(-f) > \alpha, \tag{4.3.4}$$

and from $\alpha + \lambda V f(x) + f \geq 0$,

$$\sup_{f \leq 0} V(-f) \leq \alpha$$

which contradicts (4.3.3).

Conversely, suppose (4.3.3) holds and $\alpha = \sup(\lambda V f + f)$. α is necessarily ≥ 0, since f and Vf tend to zero at infinity. $\alpha + \lambda V(-f) + (-f) \geq 0$, implying by (4.3.3), $\alpha \geq \lambda V f$. In particular, if x is a point at which $\alpha = \lambda V f(x) + f(x)$, then $f(x) \geq 0$. Thus, at every point at which $V f + f/\lambda$ attains its maximum, f is non-negative. Letting λ tend to infinity, we obtain (4.3.1).

Theorem 4.3.2. *Let V be as above. Assume that $VK(X)$ is dense in $C_0(X)$. There exists exactly one positive strongly continuous contraction semi-group P_t on $C_0(X)$ such that*

$$Vf = \int_0^\infty P_t f \, dt, \quad f \in K(X).$$

For every $f \in K(X)$, $Vf \in D(A)$ (A = the infinitesimal generator of P_t) and $AVf = -f$.

Proof. We give the proof in a series of steps. Any missing details can easily be supplied by the reader. The general idea is to define a resolvent and then use the Hille–Yosida theorem.

Step 1: For every $\lambda > 0$ and $f \in K(X)$,

$$\|\lambda V f + f\| \geq \|\lambda V f\|. \tag{4.3.5}$$

Indeed, from (4.3.3), if $\alpha = \inf(\lambda V f + f)$ and $\beta = \sup(\lambda V f + f)$, then ($\alpha \leq 0$, $\beta \geq 0$ because both $f, Vf \in C_0(X)$)

$$\alpha \leq Vf \leq \beta$$

and this implies (4.3.5).

Let $0 \leq a \leq 1$ be in $K(X)$. Define

$$V_a f = V(af), \quad f \in C_0(X).$$

From 4.3.2, if $V_a(f)$ attains strictly positive values,

$$\sup V_a f = \sup_{af > 0} V(af) = \sup_{f > 0} V_a(f),$$

i.e., V_a also satisfies the principle of positive maximum. Thus, (4.3.5) is valid with V replaced by V_a and for all $f \in C_0(X)$. Also V_a is a bounded operator since V is positive.

Step 2: For all $\lambda > 0$,

$$\text{Range}(\lambda V_a + I) = C_0(X). \tag{4.3.6}$$

That the range is closed follows from (4.3.5) with V replaced by V_a. For small λ, (4.3.6) is obvious by series expansion. Since the resolvent set of V_a is open, it is enough to show that if the claim is valid in the open interval $(0, \lambda_0)$, it is valid for λ_0. If $\lambda V_a f + f = g$, then from (4.3.5) with V replaced by V_a,

$$\|f\| \leq \|g\| + \|\lambda V_a f\| \leq \|g\| + \|\lambda V_a f + f\| = 2\|g\|,$$
$$\|\lambda_0 V_a f + f - g\| = \|\lambda_0 V_a f + f - \lambda V_a f - f\|$$
$$= |\lambda_0 - \lambda| \|V_a f\| \leq \frac{|\lambda_0 - \lambda|}{\lambda} \|g\|$$

showing that the range of $\lambda_0 V_a + I$ is dense.

Step 3: $\lambda V + I$ has dense range for all $\lambda > 0$. Let $L \in C_0(X)^*$ be such that $L(\lambda V f + f) = 0$ for all $f \in K(X)$. In particular, we have for all $0 \leq a \leq 1$, $a \in K(X)$, $L(\lambda V_a f + af) = 0$, $f \in C_0(X)$. From Step 2, for all $g \in C_0(X)$, there exists $f \in C_0(X)$ such that $\|f\| \leq 2\|g\|$ and $\lambda V_a f + f = g$. We have

$$|Lg| = |L(\lambda V_a f + f)| = |L((1-a)f)| \leq 2\|g\| |L|(1-a),$$

where $|L|$ denotes the total variation measure corresponding to L. Letting a increase to 1 leads to $L = 0$.

From Step 3 and (4.3.5), there is a bounded operator J_λ defined on $C_0(X)$ such that

$$J_\lambda (\lambda V f + f) = f, \quad f \in K(X). \tag{4.3.7}$$

Define R_λ by

$$\lambda R_\lambda = I - J_\lambda.$$

Step 4: R_λ satisfies the resolvent equation:

$$R_\lambda - R_\mu + (\lambda - \mu) R_\lambda R_\mu = 0. \tag{4.3.8}$$

Indeed, let $h \in C_0(X)$. By Step 3, there is a sequence $f_n \in K(X)$ such that $\lim (\mu V f_n + f_n) = h$. Then,

$$\lim f_n = \lim J_\mu (\mu V f_n + f_n) = J_\mu h. \tag{4.3.9}$$

Finally,

$$J_\lambda h = \lim J_\mu (\mu V f_n + f_n)$$
$$= \lim \frac{\mu}{\lambda} J_\mu (\lambda V f_n + f_n) + \frac{\lambda - \mu}{\lambda} \lim J_\lambda f_n = \frac{\mu}{\lambda} J_\mu h + \frac{\lambda - \mu}{\lambda} J_\lambda J_\mu h$$

by (4.3.7) and (4.3.9). Thus, $\lambda J_\lambda - \mu J_\mu = (\lambda - \mu) J_\lambda J_\mu$, which is equivalent to (4.3.8).

Step 5: $R_\lambda \geq 0$, $\|\lambda R_\lambda\| \leq 1$ and the range of R_λ is dense. Indeed, for $f \in K(X)$,

$$R_\lambda (\lambda V f + f) = V f \tag{4.3.10}$$

so that the range of R_λ contains the range of V. Also the same equation shows that $\|\lambda R_\lambda (\lambda V f + f)\| \leq \|\lambda V f\| \leq \|\lambda V f + f\|$, which from Step 3 is equivalent to $\|\lambda R_\lambda\| \leq 1$. That $R_\lambda \geq 0$ is equivalent to $J_\lambda h \leq h$ if $h \geq 0$. To show this, let f_n be such that $\lim (\lambda V f_n + f_n) = h$. For any $\epsilon > 0$, for all large n, $\epsilon + \lambda V f_n + f_n \geq 0$, and using (4.3.3) $\epsilon + \lambda V f_n \geq 0$ for all large n. But then, $J_\lambda h = \lim J_\lambda (\lambda V f_n + f_n) = \lim f_n \leq \lim (\epsilon + \lambda V f_n + f_n) \leq \epsilon + h$ for all $\epsilon > 0$.

An appeal to Exercise 4.2.9 gives us a strongly continuous positive semigroup P_t on $C_0(X)$ with resolvent R_λ.

Now by (4.3.10) for all $f \in K(X)$, Vf belongs to the domain $D(A)$ of generator A of P_t and

$$AVf = \lambda V f - (\lambda V f + f) = -f. \tag{4.3.11}$$

Semi-Groups of Operators, Potentials, and Diffusion Equations

From Proposition 4.3.1, $\lim \lambda R_\lambda h = 0$ for all $h \in C_0(X)$. A look at (4.3.10) then convinces us that $\lim_{\lambda \to 0} R_\lambda f = Vf$ for all $f \in K(X)$. Since P_t is non-negative, this gives (first for non-negative and then general)

$$\int_0^\infty P_t f \, dt = Vf, \quad f \in K(X). \tag{4.3.12}$$

That proves the theorem.

Example 4.3.3. Let S_λ be a sub-Markov resolvent on $\mathbb{R}^d$: The operators S_λ map the set $B(\mathbb{R}^d)$ of bounded measurable functions into itself. Note that for $S_\lambda \geq 0$, $\lambda S_\lambda \leq 1$, each S_λ is given by a measure and S_λ satisfies the resolvent equation. Suppose for a set of $f \in D(V)$,

$$V(|f|) = \lim_{\lambda \to 0} S_\lambda(|f|)$$

exists. Then for any $\epsilon > 0$, $\lambda > 0$, $f \in D(V)$,

$$\epsilon + \lambda Vf + f \geq 0 \quad \text{implies} \quad \epsilon + \lambda Vf \geq 0. \tag{4.3.13}$$

Indeed, $Vf = S_\mu f + \mu S_\mu Vf$ is valid for all $\mu > 0$ and $f \in D(V)$. Therefore, operating on the first inequality in (4.3.13), we get the second because $0 \leq \lambda S_\lambda \leq 1$.

As a special case, consider the resolvent of the Brownian semi-group. We see for $d \geq 3$ that the operator V defined on $K(\mathbb{R}^d)$ by

$$Vf(x) = \int \frac{f(y)}{|x-y|^{d-2}} dy$$

satisfies (4.3.13) and hence the principle of positive maximum.

Example 4.3.4. Let S_λ and V be as in Example 4.3.3. A non-negative Borel measurable function s is called supermedian relative to S_λ if $\lambda S_\lambda s \leq s$ for all $\lambda > 0$. Define

$$\widetilde{V}f(a) = \begin{cases} (s(a))^{-1} V(sf)(a) & s(a) \neq 0 \\ 0 & s(a) = 0. \end{cases}$$

Then V satisfies (4.3.13).

Indeed, if $\epsilon s + \lambda V(sf) + sf \geq 0$, operate by S_λ to get $\epsilon S_\lambda s + V(sf) \geq 0$ and recall that $\lambda S_\lambda s \leq s$.

Example 4.3.5. Let $0 < \alpha < 1$, $0 < T \leq \infty$. For each continuous f, define

$$u(t) = Vf(t) = \int_0^t f(t-s) s^{-\alpha} ds, \quad 0 \leq t \leq T.$$

We claim that V satisfies the principle of positive maximum in $[0, T]$. To see this, suppose first that f is continuously differentiable. Then it can be seen that

$$f(t) = \frac{1}{A} e^{-\beta} u(t) + \frac{\beta}{A} \int_0^t [u(t) - u(t-s)] s^{-1-\beta} ds,$$

where $\beta = 1 - \alpha$, $A = \int_0^1 s^{-\alpha} (1-s)^{-\beta} ds$. In particular, if t_0 is a maximum point of u in $[0, T]$, then $f(t_0) \geq A^{-1} t_0^{-\beta} u(t_0)$. [Note that $\lim_{t \to 0} t^{-\beta} u(t)$ exists.] Now approximate. The semi-group corresponding to V when $T = \infty$ is the one associated with the stable distribution with Laplace transform $\exp(-\lambda^\beta)$. See Feller [6, p. 424].

Example 4.3.6. Let again $0 < \alpha < 1$. For f bounded and integrable on $[0, \infty)$, put

$$u(t) = Vf(t) = \int_0^\infty f(t+s) s^{-\alpha} ds. \qquad (4.3.14)$$

If Vf attains strictly positive values, then

$$\sup Vf = \sup_{(f>0)} Vf. \qquad (4.3.15)$$

This can be proved as in Example 4.3.5 above. A more general procedure is the following: For each $t > 0$ there is a probability distribution F_t on $[0, \infty)$ with Laplace transform $\exp(-t\lambda^{1-\alpha})$. See Feller [6, p. 424]. The operators S_t defined by

$$S_t f(x) = \int_0^\infty f(x+y) F_t(dy)$$

form a semi-group and

$$Vf = \int_0^\infty S_t f \, dt$$

as is seen by using Laplace transforms. Thus, (4.3.13) is valid for V. For f bounded and integrable, Vf is continuous and vanishes at infinity, and, in this case, (4.3.13) can be seen to imply (4.3.15).

Semi-Groups of Operators, Potentials, and Diffusion Equations

Example 4.3.7. Let $d \geq 3$ and $0 < \alpha < 2$. For each bounded measurable function f with compact support on $\mathbb{R}^d$, define

$$I_\alpha(f)(x) = \int |x-y|^{\alpha-d} f(y)\, dy. \tag{4.3.16}$$

Then,

$$\sup I_\alpha f = \sup_{f \geq 0} I_\alpha f. \tag{4.3.17}$$

We have seen in the proof of the Riesz composition formula that $I_\alpha f$ is equal except for a constant to

$$Vf = \int_0^\infty t^{\frac{\alpha}{2}-1} T_t f\, dt, \tag{4.3.18}$$

where T_t is the Brownian semi-group. Suppose that $\epsilon + \lambda V f + f \geq 0$. We then have since $T_t \epsilon = \epsilon$,

$$\epsilon + \lambda V T_t f + T_t f \geq 0$$

for all t and x.

Fix x and let $g(t) = T_t f(x)$. Then, $VT_t f(x) = \int_0^\infty s^{\frac{\alpha}{2}-1} g(t+s)\, ds$. By Example 4.3.6,

$$\epsilon + \lambda V T_t f(x) \geq 0, \quad t \geq 0.$$

Thus, (4.3.13) is valid for V and this implies (4.3.17). The semi-group corresponding to V is the symmetric stable semi-group of exponent α and is obtained from the Brownian semi-group by the subordination procedure (Exercise 4.1.4) using the one-sided stable process on $[0, \infty)$ of exponent $\frac{\alpha}{2}$.

Example 4.3.8. Let $N(x, dy)$ be probability measures on $\mathbb{R}^d$ such that $Nf = \int N(\cdot, dy) f(y)$ is measurable for every $f \in B(\mathbb{R}^d)$ = the set of bounded measurable functions. Let $G = \sum_0^\infty N^n$, then G satisfies

$$\epsilon + Gf + f \geq 0 \quad \text{implies } \epsilon + Gf \geq 0. \tag{4.3.19}$$

Note that if the first condition in (4.3.19) is satisfied for an $f \in B(\mathbb{R}^d)$, Gf is necessarily bounded in the following. Apply N to (4.3.19) to get $\epsilon + NGf + Nf \geq 0$. Adding this to (4.3.19) and using the obvious identity $NG + I = G$,

$$2\epsilon + 2Gf + Nf \geq 0.$$

Operate by N on this last inequality and add to twice the first inequality in (4.3.19) to get

$$4\epsilon + 4Gf + N^2 f \geq 0.$$

And in general $2^n \epsilon + 2^n Gf + N^n f \geq 0$. Since f is bounded, the second inequality in (4.3.19) must be valid. The corresponding semi-group is the "Compound Poisson Semi-group":

$$S_t = \sum_0^\infty e^{-t} \frac{t^n}{n!} N^n.$$

Exercises

4.3.1. Let V satisfy the principle of positive maximum. Show that for $p > 0$, $I + pV$ also satisfies the same.

Hint: Suppose $\epsilon + \lambda (I + pV) f + f \geq 0$. Then, since V satisfies (4.3.5), $\epsilon + \lambda p V f \geq 0$. Multiply the first by $\frac{\lambda}{\lambda+1}$ and the second by $\frac{1}{\lambda+1}$ and add. The result is $\epsilon + \lambda (I + pV) f \geq 0$.

4.3.2. Let V be as above. Show that $Vf = 0$ implies $V(|f|) = 0$. In particular, for $p \geq 0$, $f + pVf = 0$ implies $f = 0$.

Hint: Let $g \in K(X)$ with support in $(|f| \geq \epsilon)$. Since $\sup Vg = \sup(Vg : g > 0)$, there is a point x_0 with $f(x_0) \geq \epsilon$ such that $Vg(x_0) = \sup Vg$. But for all α, x_0 is a maximum point of $V(g + \alpha f)$ so that $g(x_0) + \alpha f(x_0) \geq 0$. But then $f(x_0) = 0$. For the second part, use Exercise 4.3.1.

4.3.3. Let V_n be operators from $K(X)$ into $C_0(X)$, satisfying the principle of positive maximum. If $V_n f$ converges to Vf uniformly, then V satisfies the same.

References

[1] S. Al-Sharif, M. Al Horani and R. Khalil. The Hille Yosida theorem for conformable fractional semi-groups of operators. *Missouri J. Math. Sci.*, 33(1): 18–26, 2021.

[2] S. Albeverio and M. W. Yoshida. Reflection positive random fields and Dirichlet spaces. In *Proceedings of RIMS Workshop on Stochastic Analysis and Applications*, RIMS Kôkyûroku Bessatsu, B6, pp. 15–29. Research Institute for Mathematical Sciences (RIMS), Kyoto, 2008.

[3] C. Berg and G. Forst. *Potential Theory on Locally Compact Abelian Groups.* Ergebnisse der Mathematik und ihrer Grenzgebiete [Results in Mathematics and Related Areas], Band 87. Springer-Verlag, New York, 1975.

[4] A. Bobrowski. *Convergence of One-Parameter Operator Semigroups: In Models of Mathematical Biology and Elsewhere.* New Mathematical Monographs, Vol. 30. Cambridge University Press, Cambridge, 2016.

[5] J.-C. Chang and C.-L. Lang. Global existence for retarded Volterra integrodifferential equations with Hille-Yosida operators. *Dyn. Syst. Appl.*, 16(4): 625–641, 2007.

[6] W. Feller. *An Introduction to Probability Theory and Its Applications*, 2nd edn., Vol. II. John Wiley & Sons, Inc., New York, 1971.

[7] E. Hille and R. S. Phillips. *Functional Analysis and Semi-Groups.* American Mathematical Society Colloquium Publications, Vol. 31. American Mathematical Society, Providence, 1957. Revised edition.

[8] P. E. T. Jorgensen and R. T. Moore. *Operator Commutation Relations: Commutation Relations for Operators, Semigroups, and Resolvents with Applications to Mathematical Physics and Representations of Lie Groups.* Mathematics and Its Applications. D. Reidel Publishing Co., Dordrecht, 1984.

[9] P. E. T. Jorgensen, K.-H. Neeb and G. Ólafsson. Reflection positive stochastic processes indexed by Lie groups. *Symmetry Integr. Geom.: Methods Appl. (SIGMA)*, 12: 49, 2016. Paper No. 058.

[10] P. Y. G. Lamarre and M. Shkolnikov. Edge of spiked beta ensembles, stochastic Airy semigroups and reflected Brownian motions. *Ann. Inst. Henri Poincaré Probab. Stat.*, 55(3): 1402–1438, 2019.

[11] K. Matsuura. Doubly Feller property of Brownian motions with Robin boundary condition. *Potential Anal.*, 53(1): 23–53, 2020.

[12] X. L. Song, P. Zhao and X. W. Wang. Nonlinear Lipschitz perturbation of Hille-Yosida operators. *Acta Anal. Funct. Appl.*, 17(2): 130–138, 2015.

[13] M. Unser and S. Aziznejad. Convex optimization in sums of Banach spaces. *Appl. Comput. Harmon. Anal.*, 56: 1–25, 2022.

[14] K. Yosida. *Functional Analysis*, 6th edn. Grundlehren der Mathematischen Wissenschaften, Vol. 123. Springer-Verlag, Berlin, 1980.

[15] P. Zhong. Free Brownian motion and free convolution semigroups: Multiplicative case. *Pac. J. Math.*, 269(1): 219–256, 2014.

Chapter 5

Harmonic Functions, Dynkin, and Transforms

The purpose of this chapter is to offer a presentation of Harmonic functions, transforms, and potential theory with the use of the stochastic calculus introduced in the first three chapters.

The original material in Chapter 4 dealing with Harmonic functions, transforms, and potential theory is still current, and we have added a discussion of new directions. In addition, for the benefit of readers, we offer the following citations covering new directions [3, 4, 6, 9, 11, 14–16].

The new directions include sphericalization and p-harmonic functions on unbounded domains, reflectionless measures for Calderón–Zygmund operators, boundary value problems for multiplicative operator functionals of Markov processes, Weinstein operators, and Radon transforms.

For the benefit of beginner readers, we offer the following supplementary introductory text [1].

Introduction. Harmonic functions are solutions of the Laplace equation $\Delta u = 0$. No other single partial differential equation is encountered in so many different situations and exhibits such depth and variety. One runs into the Laplace equation in many branches of applied physics: electrostatics, stationary heat flow, etc. Directly or indirectly the Dirichlet problem has influenced many branches of analysis: integral equations, special functions, calculus of variations, etc. In Section 5.1, Dynkin's formula is proved and some applications are given. In Section 5.2, the Dirichlet problem is introduced. Section 5.3 deals with the Kelvin transformation. Some applications are found in the exercises and in Chapter 7. In Section 5.4, we prove the Fatou limit theorem and derive the existence of the Hilbert transform. Section 5.5, dealing with spherical harmonics, can be considered an application of the Poisson integral formula. The original idea was to give applications in representation theory but we content ourselves with a reference.

Notation. In this chapter, X_t will denote the d-dimensional Brownian motion as introduced in Chapter 3. If D is an open set, the exit time T from D is the stopping time

$$T = \begin{cases} \inf\left(t : t > 0, X_t \notin D\right) \\ \infty \quad \text{if there is no such } t. \end{cases}$$

5.1 Dynkin's Formula

If u and all its first and second partials are bounded,

$$\mathbb{E}_a\left[u\left(X_t\right)\right] - u\left(a\right) = \frac{1}{2}\mathbb{E}_a\left[\int_0^t \Delta u\left(X_s\right) ds\right], \quad a \in \mathbb{R}^d, t > 0. \quad (5.1.1)$$

The verification of (5.1.1) is simple integration by parts. The most general conditions under which (5.1.1) is valid need not concern us. For example, if u and all its first and second partials have at most polynomial growth, (5.1.1) is still valid.

Markov property and (5.1.1) imply that

$$M_t = u\left(X_t\right) - u\left(X_0\right) - \frac{1}{2}\int_0^t \Delta u\left(X_s\right) ds$$

is a martingale (relative to $\mathbb{P}_a$ for every a). Since u and Δu are bounded, M_t is bounded by a constant times t for $t \geq 1$. It is then simple to check that $\mathbb{E}\left[M_T\right] = 0$ provided $\mathbb{E}_a\left[T\right] < \infty$, where T is a stopping time. We have thus the following.

Proposition 5.1.1 (Dynkin's Formula). *Let u and all its first two partials be bounded. If T is a stopping time such that $\mathbb{E}_a\left[T\right] < \infty$, then*

$$\mathbb{E}_a\left[u\left(X_T\right)\right] - u\left(a\right) = \frac{1}{2}\mathbb{E}_a\left[\int_0^T \Delta u\left(X_s\right) ds\right]. \quad (5.1.2)$$

Suppose now that u is continuous on $\overline{D}$, Δu exists and is continuous and bounded in D, where D is a relatively compact open set. We may suppose that u in fact is continuous on $\mathbb{R}^d$ with compact support. Let $0 \leq \varphi_n$ be C^∞ and have support in $B\left(0, \frac{1}{n}\right)$ and $\int \varphi_n(x)\,dx = 1$. $u_n = u * \varphi_n$ will be C^∞ with compact support. $u_n \to u$ on $\mathbb{R}^d$ and $\Delta u_n \to \Delta u$ in D boundedly. If $T =$ exit time from D, $\mathbb{E}_a[T] < \infty$ for all a. Equation (5.1.2) is valid with u replaced by u_n and letting n tend to infinity, we get the following:

Theorem 5.1.2. *Let D be relatively compact open. If u is continuous on $\overline{D}$, Δu exists, is continuous, and is bounded in D, then for all $a \in \overline{D}$,*

$$\mathbb{E}_a[u(X_T)] - u(a) = \frac{1}{2}\mathbb{E}_a\left[\int_0^T \Delta u(X_s)\,ds\right], \qquad (5.1.3)$$

where $T =$ exit time from D.

Some consequences. A function u is called harmonic in an open set U if u is C^2 in U and $\Delta u = 0$ in U. Let us show that locally integrable function u is harmonic (in U) if and only if it has the mean value property:

$$\int_{S(a,r)} u(b)\,\mathbb{P}_a(db) = u(a), \qquad (5.1.4)$$

where $S(a,r)$ is the surface of the ball $B(a,r)$ of radius r and center a completely contained in U and $\mathbb{P}_a(db)$ is the uniform distribution on $S(a,r)$.

Let u be harmonic in U and $B(a,r)$ a ball completely contained in U. If T denotes the exit time from $B(a,r)$, we know from Chapter 3 that relative to P_a, X_T is uniformly distributed on $S(a,r)$ and $\mathbb{E}_a[T] < \infty$. Equation (5.1.4) is thus a consequence of (5.1.3). Conversely, suppose u is locally integrable and has the mean value property. If further u is C^2, Δu must be zero; because if $\Delta u(a_0) > 0$ for some a_0, in $B(a_0, r)$ for small r, Δu and for $T =$ exist time from $B(a_0, r)$, we get a contradiction using (5.1.3).

Finally, we claim that a locally integrable u having the mean value property is necessarily C^∞ in U. Indeed, let D be any relatively compact open subset of U. For r sufficiently small, $B(a,r)$ is completely contained in U for all a in $\overline{D}$. If φ is any radial (i.e., depending only on distance) C^∞-function with support in $B(0,r)$ such that $\int \varphi(b)\,db = 1$, integration using polar co-ordinates and the mean value property of u imply that $\varphi * u = u$ in D; in forming the convolution $\varphi * u$, define for definiteness

$u = 0$ off U. Since $\varphi * u$ is C^∞ in $\mathbb{R}^d$, we have shown that u is C^∞ in D, and D is any relatively compact open subset of U.

The mean value property is very useful. We have the following:

Theorem 5.1.3 (Liouville's Theorem). *A non-negative harmonic function on $\mathbb{R}^d$ is constant.*

Indeed, let a_0 be any point in $\mathbb{R}^d$. The mean value property implies that the volume average of u over any ball $B(a_0, R)$ is $u(a_0)$. From Figure 5.1, it is clear that

$u(a) =$ volume average of u over $B(a_0, R - |a_0|)$

$$\leq \frac{\text{volume } B(0, R)}{\text{volume } B(a_0, R - |a_0|)} u(0).$$

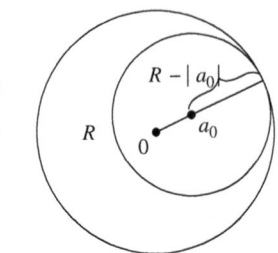

Fig. 5.1 Mean value property of a harmonic function over a spherical region.

Letting R tend to infinity, we find $u(a) \leq u(0)$. For a more general statement, see Exercise 5.1.4.

Theorem 5.1.4 (Harnack's Theorem). *Let U_n be harmonic and increase in a connected open set D. Then $u = \lim_n u_n$ is harmonic in D unless it is identically infinite.*

Indeed, by considering $u_n - u_0$ instead of u_n, we may assume $u_n \geq 0$. u satisfies the mean value property. If $u(a_0) < \infty$, u is integrable in any ball $B(a_0, r)$ completely contained in D. But then the mean value property would imply that u is harmonic and hence finite in such a ball. By connectedness, for any a in D, there is a finite chain of intersecting balls, the first containing a_0 and the last containing a.

Exercise 5.1.5 gives an example of a non-harmonic function having the one-circle mean value property. See also Exercise 5.1.7.

An immediate consequence of (5.1.3) is as follows: If u is continuous on $\overline{D}$ and harmonic in a bounded open set D, then

$$u(a) = \mathbb{E}_a[u(X_T)], \quad a \in \overline{D}, \tag{5.1.5}$$

where $T =$ exit time from D. Since $X_T \in \partial D$, the boundary of D, we obtain the following:

Theorem 5.1.5 (The Maximum Principle). *If u is harmonic in a bounded open set D and continuous on $\overline{D}$, then*

$$|u(a)| \leq \sup_{b \in \partial D} |u(b)|, \quad a \in \overline{D}. \tag{5.1.6}$$

Let us work out another consequence of (5.1.5). Direct calculation shows that for $d \geq 3$, $u(a) = |a|^{-d+2}$ is harmonic in the complement of $\{0\}$. Take in (5.1.5) for D the domain bounded by two concentric spheres: $D = \{a : r < |a| < R\}$. Since $|X_T|$ is either R or r and $\mathbb{P}_a[|X_T| = R] = 1 - \mathbb{P}_a[|X_T| = r]$, we get for $d \geq 3$, $r < |a| < R$,

$$\mathbb{P}_a[|X_T| = r] = \frac{|a|^{-d+2} - R^{-d+2}}{r^{-d+2} - R^{-d+2}}, \quad d \geq 3, \tag{5.1.7}$$

where T = exit time from D. When $d = 2$, $u(a) = \log|a|$ is harmonic in the complement of $\{0\}$. By the same argument as above,

$$\mathbb{P}_a[|X_T| = r] = \frac{\log R - \log|a|}{\log R - \log r}, \quad d = 2. \tag{5.1.8}$$

The event $(|X_T| = 1)$ occurs if and only if at the time T of exit from the shell $(r < |x| < R)$ the Brownian path finds itself on the sphere $S(0, r) = (x : |x| = r)$. Letting R tend to infinity, we find from (5.1.7) and (5.1.8) that for $|a| > r$,

$$\mathbb{P}_a[T_r < \infty] = \begin{cases} 1 & d = 2 \\ r^{d-2}|a|^{2-d} & d \geq 3, \end{cases} \tag{5.1.9}$$

where T_r is the hitting time to $S(0, r)$:

$$T_r = \begin{cases} \inf(t : t > 0, |X_t| = r) \\ \infty \quad \text{if there is no such } t. \end{cases}$$

The above results imply that the d-dimensional Brownian motion is "recurrent" if $d \leq 2$ and "transient" if $d \geq 3$; if $d \geq 2$, $\mathbb{P}_a[T_0 < \infty] = 0$ for all a, where T_0 is the hitting time to zero: $T_0 = \inf(t : t > 0, X_t = 0)$. See Exercises 5.1.2 and 5.1.3.

Exercises

5.1.1. Let T = exit time from the ball $B(a_0, r)$. Show that

$$d\mathbb{E}_a[T] = r^2 - |a - a_0|^2.$$

Hint: Take $u(a) = |a - a_0|^2$ in (5.1.5).

5.1.2. Let $d \geq 2$ and $T_0 =$ hitting time to 0. Show that

$$\mathbb{P}_a [T_0 < \infty] = 0 \quad \text{for all } a.$$

Hint: Suppose $a \neq 0$. Let $r \to 0$ in (5.1.7) and (5.1.8) and then let R tend to infinity. One obtains $\mathbb{P}_a [T_0 < \infty] = 0$ for all $a \neq 0$. Deduce using the Markov property that $\mathbb{P}_0 [T_0 (\theta_t) < \infty]$ for all $t > 0$. Conclude that $\mathbb{P}_0 [T_0 < \infty] = 0$.

5.1.3. Show that the Brownian motion is recurrent for $d = 2$ and transient for $d \geq 3$: For all a,

$$\mathbb{P}_a \left[\lim_{t \to \infty} |X_t| = \infty \right] = 1, \quad \text{if } d \geq 3,$$

and

$$\mathbb{P}_a \left[(X_s : s \geq t) \text{ is dense in } \mathbb{R}^2 \text{ for all } t \right] = 1, \quad \text{if } d = 2.$$

Hint: Define T_n, τ_n by

$$T_n = \inf \left(t : t > 0, |X_t| \leq n \right),$$
$$\tau_n = \inf \left(t : t > 0, |X_t| \geq n^3 \right)$$

with the convention that the infimum over an empty set is ∞. For all n, $\mathbb{P}_a [\tau_n < \infty] = 1$ (Exercise 5.1.1). Use the strong Markov property and (5.1.9):

$$\mathbb{P}_a [T_n (\theta_{\tau_n}) < \infty] = \begin{cases} n^{-2d+4} & d \geq 3 \\ 1 & d = 2. \end{cases}$$

Thus, for $d \geq 3$,

$$\sum_1^\infty \mathbb{P}_a \left[|X(t + \tau_n)| \leq n \text{ for some } t \right] = \sum_1^\infty \mathbb{P}_a \left[|T_n (\theta_{\tau_n})| < \infty \right].$$

Now use Borel–Cantelli lemma.

5.1.4. Let u be harmonic on $\mathbb{R}^d$ and $u(x) \leq \alpha |x| + \beta$ for all x, where α and β are constants. Then, u is affine, i.e., $u - u(0)$ is linear.

Hint: Let $x \in \mathbb{R}^d$ be fixed. Let A and $A(t)$ denote the balls $B(x, R)$ and $B(x + th, R)$, where h is a unit vector. Using spatial mean values ($|B|$

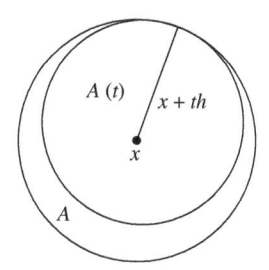

Fig. 5.2 Behavior of harmonic functions using spatial mean values along a direction.

denoting volume of B) (Figure 5.2),

$$u(x+th) - u(x) = \left(\frac{1}{|A(t)|} - \frac{1}{|A|}\right)\int_{A(t)} u - \frac{1}{|A|}\int_{A\setminus A(t)} u,$$

which, using $u(y) \le \alpha |y| + \beta$, shows that the derivatives of u in the direction h are at least equal to $\frac{u(x)}{R} - \frac{\alpha|x|+\alpha R+\beta}{R}$. Let R tend to infinity to conclude, using Liouville's theorem (Theorem 5.1.3) that $\frac{\partial u}{\partial h}$ is a constant.

5.1.5. The Bessel function $J_0(\xi)$ of order 0 may be defined by

$$J_0(\xi) = \frac{1}{2\pi}\int_0^{2\pi} e^{-i\xi\sin\theta}d\theta.$$

Show that if $\xi = a+ib$ is a root of $J_0(z)$ and $u(x,y) = \Re e^{i\xi y} = e^{-by}\cos ay$, then u has the "one circle mean value property", i.e., for each $z_0 = (x_0, y_0)$,

$$u(z_0) = \int_{S(z_0,1)} u(z)\,\sigma(dz),$$

where $S(z_0, 1)$ is the circle of radius 1 center z_0. u is not harmonic unless $\xi = a + ib = 0$.

Hint: That u has the one circle mean value property follows from the definition. Since u is independent of x, it cannot be harmonic unless it is a constant. That $J_0(\xi) = 1$ has infinitely many roots follows from Exercise 5.1.6.

5.1.6. Let f be entire and $|f(z)| \le e^{A|z|}$ for all large enough $|z|$ for some constant A. If f has only finitely many zeroes, then $f(z) = P(z)e^{az}$ for some a and polynomial P. In particular, f entire, $|f(z)| \le e^{A|z|}$, f bounded on the real axis implies $f(z) - a = 0$ has infinitely many roots for every complex a for which the equation has at least one root.

Hint: If f has only finitely many zeroes, we can write $f(z) = p(z) e^{G(z)}$, where G is entire and p is a polynomial. If $G = u + iv$, the conditions on f imply $u(z) \leq A|z|$ for all large $|z|$. By Exercise 5.1.4, u is linear and this implies $G(z) = \alpha z$. If f is bounded on the real axis, p must reduce to a constant and α must be purely imaginary, i.e., $f(z) = A e^{i\alpha z}$, α real. But then f never vanishes. For any a, we can apply the above reasoning to $f(z) - a$ to conclude that if $f(z) - a = 0$ has one solution, it has infinitely many. Since $J_0(0) = 1$ and $|J_0(x)| \leq 1$ for x real, we conclude $J_0(x) = 1$ has infinitely many roots.

Theorem 5.2.2 is needed for Exercise 5.1.7.

5.1.7. Let D be a bounded domain in $\mathbb{R}^d$ for which every point of ∂D is regular. If u is continuous on D, for each $x \in D$, there is a ball (depending on x), $B(x, r)$ such that $u(x) = \int_{S(x,r)} u(z) \sigma(dz)$, then u is harmonic in D.

Hint: Let v be harmonic with boundary values $u|_{\partial D}$. Then v is continuous on $\overline{D}$; $w = u - v$ is continuous on $\overline{D}$ and vanishes on ∂D. Since for each x there is a ball such that the mean value over it of w is equal to $w(x)$, w cannot attain its maximum or minimum inside D. This implies $w \equiv 0$.

5.2 Dirichlet Problem

Given a domain D and a continuous function f on the boundary ∂D, the Dirichlet problem consists in finding a function u which is continuous on $\partial \overline{D}$ and harmonic in D such that $u = f$ on ∂D. If D is bounded, we know from the maximum principle that there can be at most one such harmonic function. In other words, if D is bounded, the solution of the Dirichlet problem is unique if it exists.

The Dirichlet problem as stated above does not always have a solution. We soon give necessary and sufficient conditions for a solution to exist. These conditions involve individual points in the boundary ∂D. Roughly speaking, the complement of D should not be too small in any neighborhood of any point of ∂D.

Let T denote the exit time from D (this is the same as saying T is the hitting time to the complement of D):

$$T = \inf(t : t > 0, X_t \notin D),$$

the *infimum over an empty set is always by definition* ∞. T is a Markov time. From the zero-one law in Chapter 3, for any a, $\mathbb{P}_a(T = 0) = 1$ or 0.

Definition 5.2.1. A point $a \in \partial D$ is called *regular* for $D^c =$ the complement of D if $\mathbb{P}_a(T = 0) = 1$. Otherwise, it is called *irregular*.

Starting at a regular point, the Brownian path hits the complement immediately. In this sense, the complement of D is not too thin at any regular point. We have the following:

Theorem 5.2.2. *Let D be a bounded open set. The Dirichlet problem is solvable for D for all continuous f on ∂D if and only if every point of ∂D is regular.*

For the proof, we need the following.

Proposition 5.2.3. *Let D be an open set and $T =$ exit time from D. For each $t > 0$, the function*

$$\mathbb{P}_a(T \geq t)$$

is upper semicontinuous on $\mathbb{R}^d$.

Proof. It is easy to show that the stopping times $s + T(\theta_s)$ decrease to T as s decreases to zero. Thus,

$$\mathbb{P}_a(T \geq t) = \inf_{s>0} \mathbb{P}_a(s + T(\theta_s) \geq t).$$

The function $\mathbb{P}_a(s + T(\theta_s) \geq t) = \mathbb{E}_a(\mathbb{P}_{X_s}(T \geq t - s))$ is continuous on $\mathbb{R}^d$ for each $s > 0$ because $\mathbb{P}_{X_s}(T \geq t - s)$ is bounded and measurable in a. $\square$

Proof of Theorem 5.2.2. Suppose every point in ∂D is regular and f is any continuous function on ∂D. Let

$$u(a) = \mathbb{E}_a(f(X_T)), \quad a \in \overline{D},$$

where $T =$ exit time from D. Clearly, $u = f$ on ∂D. If $a \in D$, $B(a,r)$ a ball contained in D, and $S =$ exit time from $B(a,r)$, then $T = S + T(\theta_S)$ and using the strong Markov property,

$$u(a) = \mathbb{E}_a(u(X_S)).$$

Since X_S is uniformly distributed on $\partial B(a,r)$ relative to $\mathbb{P}_a$, u has the mean value property in D. To check the continuity of u at a point $a_0 \in \partial D$, we need only show that for a close enough to a_0, X_T is close to a_0 or just that X_T is close to a with large $\mathbb{P}_a$-probability. Now given $\epsilon > 0$,

$$\mathbb{P}_a \left(\sup_{0 \leq s \leq t} |X_s - a| \geq \epsilon \right) \tag{5.2.1}$$

is uniformly small provided t is small enough (Exercise 5.2.1). For this t, by Proposition 5.2.3, $\mathbb{P}_a(T \geq t)$ is small when a is close to a_0. By (5.2.1), the $\mathbb{P}_a$-probability that $T \leq t$ and $|X_T - a| \geq \epsilon$ is small. Thus, X_T is close to a (with large $\mathbb{P}_a$-probability) if a is close to a_0.

Conversely, suppose that the Dirichlet problem is solvable for D for all continuous boundary data. Let $a_0 \in \partial D$ and f be any non-negative continuous function on ∂D which vanishes exactly at a_0. By assumption, there is a continuous function u on $\overline{D}$ which is harmonic in D and whose restriction to ∂D is f. From (5.1.5), we have

$$u(a) = \mathbb{E}_a(f(X_T)), \quad a \in \overline{D}. \tag{5.2.2}$$

Taking $a = a_0$ in (5.2.2), we conclude $\mathbb{P}_a(T = 0) = 1$.

In view of the above theorem, it is natural to seek conditions for regularity of boundary points. But first a few remarks. □

Remark 5.2.4. In the first part of the proof of the above theorem, we have in fact shown the following: For any open set D and any bounded continuous function f on ∂D, the function u defined on $\overline{D}$ by $u(a) = \mathbb{E}_a(f(X_T) : T < \infty)$ is harmonic in D and continuous at every regular point in ∂D. Of course, $u(a) = f(a)$ if $a \in \partial D$ is regular.

Our second remark is that regularity is a local property: A point $a \in \partial D$ is regular for D^c if and only if it is regular for $(D \cap U)^c$, where U is any open neighborhood of a. Indeed, if a is irregular, the Brownian path starting at a remains in D for a short length of time so that it remains in $D \cap U$ for a short length of time. And conversely. Exercise 5.2.4 claims that the set of regular points is a non-empty G_δ-subset of ∂D.

The following is a nice geometrical condition guaranteeing regularity. It is called the Poincare cone condition.

Proposition 5.2.5. *Let D be open and $a \in \partial D$. If there is a cone with vertex a contained in D^c, then a is regular.*

Proof. If $B(a, r)$ is the ball with center a and radius r, and T = exit time from $B(a, r)$, $\mathbb{P}_a(X(T_r) \in D^c \cap S(a, r))$ = the uniform measure of $D^c \cap S(a, r) \geq$ the uniform measure of the part of $S(a, r)$ in the cone (Figure 5.3), and this measure is independent of r. Now T = the exit time from D clearly cannot exceed T_r on the set $X(T_r) \in D^c$. Thus, $\mathbb{P}_a(T \leq T_r)$ is bounded below by a fixed constant. Letting r tend to zero, we see that $\mathbb{P}_a(T = 0) > 0$. □

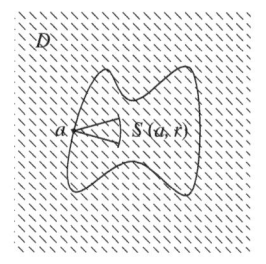

Fig. 5.3 Cone condition ensuring regularity at boundary point.

The above cone condition shows that if the boundary of an open set D is piecewise smooth, then every point of ∂D is regular. In particular, the Dirichlet problem is solvable for open sets with piecewise smooth boundaries.

The union of two open sets with regular boundaries need not have regular boundary: It is easy to find finite number of balls whose union is the unit ball punctured at the origin. However, every open set is easily seen to be an increasing union of open sets with piecewise smooth boundaries.

Since the Dirichlet problem is not always solvable, Wiener formulated the so-called modified Dirichlet problem: Given a bounded open set D and a continuous function f on ∂D, find a function u which harmonic in D and

$$\lim_{D \ni u \to b} u(x) = f(b)$$

for every $b \in \partial D$ that is regular for D^c.

We have seen (Remark 5.2.4) that such a function always exists. Wiener established the existence of a solution by a limit procedure: He wrote the open set as the increasing union of open sets with smooth boundaries and showed that solutions for these open sets converged to a solution for the union. The problem of uniqueness seems to have defeated Wiener. The maximum principle cannot be applied and a much more delicate technique is needed. O. D. Kellog established uniqueness for $d = 2$ and G. Evans for $d \geq 3$. We return to this in Chapter 6.

Example 5.2.6 (Poisson Integral Formula). Direct calculation shows that u defined by

$$u(b) = \int_{S(a,r)} P_r(b,z) f(z) p_r(dz) \tag{5.2.3}$$

solves the Dirichlet problem for the ball $B(a,r)$ with boundary data a continuous function f on $S(a,r)$. Here the Poisson kernel $P_r(b,z)$ is

$$P_r(b,z) = r^{d-2} \frac{r^2 - |b-a|^2}{|b-z|^d}, \quad d \geq 2, \tag{5.2.4}$$

and $p_r(dz)$ is the uniform distribution on $S(a,r)$. See Exercise 5.2.6 for a verification. The maximum principle of course implies that a function u which is continuous on $B(a,r)$ and harmonic in its interior is given by (5.2.3).

Formula (5.2.3) is very useful. As an illustration, let us show that u is positive and harmonic in open ball $B(0,1)$ if and only if

$$u(b) = \int P(b,z) m(dz), \quad |b| < 1 \tag{5.2.5}$$

for a unique positive bounded measure m on $S(0,1)$. Here $p(b,z)$ is given by (5.2.4) with $r=1$ and $a=0$. Indeed, for $t<1$, the mean value property applied to $B(0,t)$ shows that $u(z)p_t(dz)$ are uniformly bounded measures on the closed ball $B(0,1)$, $p_t(dz)$ being the uniform measure on $S(0,t)$. Let m denote any weak limit. Apply formula (5.2.3) for $B(0,t)$ and let t tend to 1 to get the above representation (5.2.5). For uniqueness, see Exercise 5.2.12. For other applications on the formula, see Exercises 5.2.8 and 5.2.11.

Poisson formula for the half-space. If f is bounded and continuous on $\mathbb{R}^d$,

$$u(x,t) = \Gamma\left(\frac{d+1}{2}\right) \pi^{-\frac{d+1}{2}} \int_{\mathbb{R}^d} \frac{t}{\left(|x-y|^2 + t^2\right)^{\frac{d+1}{2}}} f(y) \, dy \tag{5.2.6}$$

is harmonic in the half space $\{(x,t) : x \in \mathbb{R}^d, t > 0\}$, continuous on its closure, and equals f on its boundary. The constant in (5.2.6) is chosen so that $u=1$ when $f=1$. The verification is similar to that in Exercise 5.2.6. For a "derivation" of formula (5.2.6), see Exercise 5.2.7.

If we replace $f(z)p(dz)$ and $f(y)dy$ by $m(dz)$, where m is positive, in (5.2.3) and (5.2.6), the resulting functions are harmonic in the ball and half space respectively unless they are identically infinite. This is easy to see. Harnack's theorem in Section 5.1 shows that if D is a connected open set and f is non-negative measurable on ∂D, then u defined in D by $u(a) = \mathbb{E}_a[f(X_T) : T < \infty]$ is harmonic in D provided it is finite at one point; T is as usual the exit time from D.

For any constant c, $ct + u(x,t)$ with u defined by (5.2.6) is harmonic in the half space $\{(x,t) : x \in \mathbb{R}^d, t > 0\}$ and assumes the same boundary values as u, namely f. However, the only bounded solution is given by (5.2.6); we shall see this in Corollary 5.3.3. Granting this consider the following application: For all $a \in \mathbb{R}^d$, the function $\exp(-|\alpha|t + i\alpha \cdot x)$ is harmonic on $\mathbb{R}^{d+1}$ and in particular on the half space $\{(x,t) : t > 0\}$ with boundary values $e^{i\alpha \cdot x}$; here $\alpha \cdot x$ denotes inner product. We must have

$$\exp(-|\alpha|t + i\alpha \cdot x) = \Gamma\left(\frac{d+1}{2}\right) \pi^{-\frac{d+1}{2}} \int_{\mathbb{R}^d} \frac{te^{i\alpha \cdot y}}{\left(|x-y|^2 + t^2\right)^{\frac{d+1}{2}}} dy.$$

Taking $x = 0$, we see that the probability measures

$$F_t(dy) = \pi^{-\frac{d+1}{2}} \Gamma\left(\frac{d+1}{2}\right) t \left(|y|^2 + t^2\right)^{-\frac{d+1}{2}} dy$$

on $\mathbb{R}^d$ form a semi-group (under convolution). These are the so-called symmetric Cauchy distributions on $\mathbb{R}^d$. We also see that the Fourier transform of F_t is $e^{-|\alpha|t}$. See Feller [8, pp. 69–72].

As another application, let z_k be an infinite sequence of complex numbers with $\Im z_k \neq 0$. Assume that $\lim z_k = z_0$ exists and $\Im z_0 \neq 0$. Let S be the set of functions on $\mathbb{R}^1$, $S = \left(\frac{1}{x - z_k}, \frac{1}{x - \bar{z}_k}, k \geq 1\right)$. We claim that the linear span of S is dense (with regard to the uniform norm) in the space of all complex continuous functions on $\mathbb{R}^1$ which vanish at ∞. To show this, let m be any real bounded measure for which

$$\int \frac{m(dx)}{x - z_k} = 0, \quad k \geq 1.$$

The function $F(z) = \int \frac{m(dx)}{x - z_k}$ is holomorphic on $\Im(z) > 0$. Assuming $\Im z_0 > 0$, otherwise look at $\bar{z}_k$, $F(z) = 0$ for an infinite set of values z with limit point, namely the set $\{z_k, z_0\}$. It follows that $F \equiv 0$. Taking real parts,

$$\int \frac{t}{(x-y)^2 + t^2} m(dt) = 0, \quad t > 0.$$

This implies $m = 0$. Indeed, if f is continuous and bounded, using Fubini,

$$\int m(dy) \int \frac{t}{(x-y)^2 + t^2} f(x)\, dx = 0, \quad t > 0.$$

As t tends to zero, the inner integral tends to $\pi f(y)$ proving that $\int f(y)\, m(dy) = 0$, i.e., $m = 0$.

Example 5.2.7. Coordinate maps on $\mathbb{R}^d$ are clearly harmonic. In $\mathbb{R}^1$, the only harmonic functions are linear. The real part of any holomorphic function on $\mathbb{R}^2$ is harmonic, so is the logarithm of its modulus provided it does not vanish. In general, solutions of the Dirichlet problem are given in series using the method of separation of variables as the following example illustrates.

Example 5.2.8. The function $\exp\left(\pm |n| y\right) \sin nx\, x$ is harmonic on
$$\mathbb{R}^d = \left\{(x,y) : x \in \mathbb{R}^{d-1},\, y \in \mathbb{R}^1\right\}.$$

Here, $n = (n_1, \ldots, n_{d-1})$, $x = (x_1, \ldots, x_{n-1})$, $\sin nx = \prod_1^{d-1} \sin n_i x_i$, and $|n|^2 = \sum n_i^2$.

Superposition of such functions solves the Dirichlet problem for the cube. Thus,
$$u(x,y) = \sum \left(A_n e^{-|n|y} + B_n e^{ny}\right) a_n \sin nx$$

is harmonic in the domain $\{(x,y) : 0 < x_i < \pi,\, 0 < y < 1\}$ and assumes the value 0 on the boundary except on the side $(y = 0)$ where it assumes the value $h(x)$; A_n, B_n are found from the following: $A_n + B_n = 1$, $A_n e^{-|n|} + B_n e^{|n|} = 0$ and a_n are the Fourier coefficients of h:
$$h(x) = \sum a_n \sin nx.$$

Example 5.2.9. A very trivial example of an irregular boundary point is obtained when we remove zero from the plane. A more serious example is the following:

Let $D = B(0,1) \setminus \{x \in \mathbb{R}^3 : x_1 \geq 0,\, x_2 = x_3 = 0\}$, i.e., D is the domain obtained by removing the non-negative x_1-axis from the unit ball in $\mathbb{R}^3$. See Figure 5.4.

We claim that 0 (in fact, every point $(x_1, 0, 0)$, $0 \leq x_1 < 1$) is irregular for D^c. Indeed, writing the three-dimensional Brownian motion $X(t) = (X_1(t), X_2(t), X_3(t))$,

$$P_0(X_t \in x_1\text{-axis for some } t)$$
$$= P_{(0,0)}((X_2(t), X_3(t)) = (0,0) \text{ for some } t) = 0$$

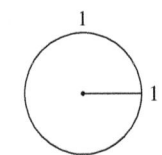

Fig. 5.4 Irregular boundary in 3D with the x_1-axis removed from the unit ball.

since $(X_2(t), X_3(t))$ is the two-dimensional Brownian motion, and, by Exercise 5.1.2, it never hits any point.

The same example works in any dimension $d \geq 3$. The case $d = 2$ is very different and much harder. Here is an example.

Harmonic Functions, Dynkin, and Transforms

From the open unit disc in the plane, we remove small discs along the x-axis so that for the resulting open set O becomes irregular (Figure 5.5). To see how this can be done, note that the probability of hitting, starting at 0, a disc $B(a,r)$ $(r < |a|)$ before exiting from the unit disc is, since $B(a, |a|+1)$ contains the unit disc, less or equal to the probability, starting at 0, of hitting $B(a,r)$ before exiting from $B(a, |a|+1)$. This last is, by (5.1.8),

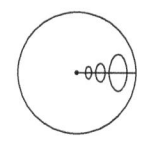

Fig. 5.5 Irregular boundary formed by removing small discs from an open set.

$$\frac{\log(1+|a|) - \log|a|}{\log(1+|a|) - \log r}.$$

Let a_n be a sequence, $a_n \downarrow 0$, $r_n < a_n$, so that say, $\left[\frac{1}{a_n}+1\right]^{2^n} \le \frac{1}{r_n} + \frac{a_n}{r_n}$. Let $D = B(0,1) \setminus \{\bigcup_n B(a_n, r_n) \cup \{0\}\}$. From the above, we see

$$\sum_n \mathbb{P}_0 \left[\text{hitting } B(a_n, r_n) \text{ before exiting from } B(0,1)\right] < \infty.$$

That is, with $\mathbb{P}_0$-probability 1, the Brownian path hits only finitely many $B(a_n, r_n)$ before hitting $\{x : |x| = 1\}$. This clearly means that 0 is irregular for D^c.

The above is also an example of an irregular point whose "harmonic measure has infinite energy". We return to this in Chapter 8.

Exercises

5.2.1. Show that $\frac{1}{t}\mathbb{P}_a\left(\sup_{0\le s \le t} |X_s - a| \ge \epsilon\right)$ tends uniformly to zero as t tends to zero.

Hint: $\mathbb{P}_a\left(\sup_{0\le s \le t} |X_s - a| \ge \epsilon\right) = \mathbb{P}_0\left(\sup_{0 \le s \le t} |X_s| \ge \epsilon\right)$. Now consider the fourth moment of X_t.

5.2.2. Let D be open and T = exist time from D. For every compact $K \subset D$, $\lim_{t \to 0} \sup_{a \in K} \mathbb{P}_a(T \le t) = 0$.

Hint: Let ϵ = distance of K to the boundary of D in Exercise 5.2.1.

5.2.3. Let D and T be as in Exercise 5.2.2. Show that for each bounded Borel measurable f on D and $t > 0$, $\mathbb{E}_a(f(X_t) : t < T)$ is a continuous function of a in D.

Hint: $T = s + T(\theta_s)$ on the set $T > s$. So for all $s < t$,

$$\mathbb{E}_a(f(X_t) : t < T)$$

$$= \mathbb{E}_a \left[\mathbb{E}_{X_s} \left(f(X_{t-s}) : t-s < T \right) : s < T \right]$$
$$= \mathbb{E}_a \left[\mathbb{E}_{X_s} \left(f(X_{t-s}) : t-s < T \right) \right]$$
$$- \mathbb{E}_a \left[\mathbb{E}_{X_s} \left(f(X_{t-s}) : t-s < T \right) : T \leq s \right].$$

The first term is continuous on $\mathbb{R}^d$ and the last term tends uniformly to zero on compact subsets of D from Exercise 5.2.2.

5.2.4. Let D be an open set. Show that the set of regular points in ∂D is a non-empty G_δ-subset.

Hint: The set $\bigcap_n \{a : a \in \partial D, \mathbb{P}_a (T < \frac{1}{n}) > 1 - \frac{1}{n}\}$ is the set of regular points. Each set of the intersection is open in ∂D by Proposition 5.2.3. To establish non-emptiness, we may assume D is bounded. Let T, S be the exit times from D and $\overline{D}$ respectively and $a \in D$. $T \leq S$ and $S(\theta_S) = 0$ if $S < \infty$ so that every point X_S is regular for D, $\mathbb{P}_a$-almost surely.

5.2.5. Let D be any open set and $a \in \partial D$ be irregular. Then a_0 is in the boundary of a connected component of D.

Hint: Starting at a_0, the continuous Brownian path is in D for a short length of time.

5.2.6. Verify that the Poisson integral formula does indeed solve the Dirichlet problem for the ball.

Hint: Assume $a = 0$. For fixed z, $|b-z|^{-d+2}$ is harmonic in b except at z. By differentiation, $(z_i - b_i)|b-z|^{-d}$ is harmonic except at z. Multiplying by $2z_i$ and adding, we get the following: $\left(2|z|^2 - 2(z,b) \right) |b-z|^{-d}$ is harmonic except at z. Now subtract $|b-z|^{-d}$ to see that $\left(|z|^2 - |b|^2 \right) |b-z|^{-d}$ is harmonic except at z. Thus, u defined by (5.2.3) is certainly harmonic in $B(0,r)$. Improving the continuity of u on the closed ball, the only non-routine verification is that $u = 1$ if $f = 1$. This is verified by noting that u is then rotation invariant and that a function which is harmonic in a neighborhood of 0 and which depends only on distance is a constant: $\Delta u = \frac{d^2 u}{dr^2} + \frac{d-1}{r} \frac{du}{dr}$ if $u(a) = u(|a|)$, $|a| = 1$.

5.2.7. Let D be the half-space $\{(a,b) : a \in \mathbb{R}^{d-1}, b > 0\}$ and $T =$ exit time from D. Show that the distribution of X_T relative to $\mathbb{P}_{a,b}$ has density

$$\pi^{-1/2} \Gamma(d/2) b \left(|a-x|^2 + b^2 \right)^{-d/2}.$$

Hint: Write the d-dimensional Brownian motion as (X_t, Y_t), where X_t is $(d-1)$-dimensional and Y_t one-dimensional. The exit time T from the half-space is simply inf $(t: t > 0, Y_t = 0)$. Relative to $\mathbb{P}_{(a,b)}$, the processes $(X\cdot)$, $(T\cdot)$ are independent. In Chapter 3, Section 3.2, the distribution of T has been found using the first passage time relation. Thus, the variable (X_T, Y_T) has, relative to $\mathbb{P}_{(a,b)}$, the $(d-1)$-dimensional density

$$\int_0^\infty \frac{1}{(2\pi t)^{\frac{d}{2}-1}} \exp\left(-\frac{|x-a|^2}{2t}\right) \frac{1}{\sqrt{2\pi}} \int_0^b \left(\frac{1}{t^{3/2}} - \frac{y^2}{t^{5/2}}\right) e^{-y^2/2t} dy dt.$$

Integrate by parts the integral $\int_0^b \exp\left(-\frac{|x-a|^2+y^2}{2t}\right) dy$ to see that the required density is equal to

$$b(2\pi)^{-d/2} \int_0^\infty t^{-\frac{d+2}{2}} \exp\left(-\frac{|x-a|^2+y^2}{2t}\right) dt$$

which is easily evaluated.

5.2.8 (Harnack's Inequalities). Let u be positive and harmonic in $B(0,r)$. Then for $|a| < r$,

$$r^{d-2} \frac{r^2 - |a|^2}{(r+|a|)^d} u(0) \leq u(a) \leq r^{d-2} \frac{r^2 - |a|^2}{(r-|a|)^d} u(0).$$

Hint: Use the Poisson integral formula.

5.2.9 (Generalized Harnack Inequalities). Let D be connected and open and K a compact subset of D. There exists a number M depending only on K and D such that for all positive harmonic functions u on D,

$$\frac{u(a)}{u(b)} \leq M, \quad a, b \in K.$$

Hint: Use Exercise 5.2.7 above and cover K suitably by balls.

5.2.10. Let u_n be positive harmonic functions in a connected open set D. If for some $x_0 \in D$, $\lim u_n(x_0) = 0$, then u_n converges to zero uniformly on compact subsets of D.

Hint: Use Exercise 5.2.9.

5.2.11. Show that a harmonic function is real analytic.

Hint: Use Poisson integral formula.

5.2.12. If for a signed measure m on $S(0,1)$,

$$v(b) = \int P(b,z) m(dz), \quad |b| < 1,$$

then $m = 0$. $P(b,z)$ is defined by (5.2.4) with $a = 0$ and $r = 1$.

Hint: If f is continuous on $S(0,1)$ and $u(a) = \int P(a,z) f(z) p(dz)$, p = uniform measure on $S(0,1)$, then $\lim_{r \to 1} u(ra) = f(a)$ boundedly for all $a \in S(0,1)$. Now use Fubini, dominated convergence and $p(a, rz) = P(ra, z)$ to get

$$\int f(a) m(da) = \lim_{r \to 1} \int f(z) p(dz) v(rz) = 0.$$

5.2.13. Let u be harmonic in the ball $B(0,1)$. u is the difference of two non-negative harmonic functions v and w in $B(0,1)$ if and only if

$$\sup_{r<1} \int |u(z)| \, p_r(dz) < \infty,$$

where $p_r(dz)$ = uniform distribution on $S(0,1)$.

Hint: Let m be the weak limit of $|u(z)| p_r(dz)$ as r tends to 1, there can only be one by Exercise 5.2.12. $P_r(b,z)$ tends uniformly to $P(b,z)$ as r tends to 1. Use this to show that

$$|u(b)| \leq \int_{S(0,1)} P(b,z) m(dz).$$

5.2.14. Let D be an open set. A uniformly bounded family of harmonic functions on D is equicontinuous on every compact subset K, i.e., given $\epsilon > 0$, there exists a $\delta > 0$ such that $|u(a) - u(b)| \leq \epsilon$ for all $a, b \in K$, $|a - b| < \delta$.

Hint: Use the Poisson integral formula and the boundedness of the family to see that the family must be locally equicontinuous.

5.3 The Kelvin Transformation

Let $B(a, \rho)$ be the ball with center a and radius ρ. For $x \in \mathbb{R}^d \setminus \{a\}$, the point x^* defined by

$$x^* = a + \frac{\rho^2}{|x-a|^2}(x - a)$$

is called the *inverse* of x relative to $\partial B(a, \rho) = S(a, \rho)$. x^* lies on the line joining a and x and $|x^* - a| \cdot |x - a| = \rho^2$ (Figure 5.6). The map $x \mapsto x^*$ is a homomorphism on $\mathbb{R}^d \setminus \{a\}$ onto $\mathbb{R}^d \setminus \{a\}$.

If f and g are related by $g(x) = f(x^*)$ elementary calculations, show that

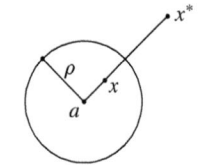

Fig. 5.6 Inversion map in a ball, showing harmonicity under Kelvin transformation.

$$\Delta g(x) = \frac{4}{r^4}(\Delta f)(x^*) + \frac{2\rho^2}{r^4}(2-d) \sum_j \frac{\partial f}{\partial x_j}(x^*)(x_j - a_j),$$

where $r = |a - x|$. Thus, if $d = 2$, the harmonicity of f in a neighborhood of x^* implies that of g in a neighborhood of x. For $d > 2$, this is false. However, the function $\frac{\rho^{d-2}}{r^{d-2}} g$ is harmonic in a neighborhood of x if f is harmonic in a neighborhood of x^*. The map $f \mapsto \frac{\rho^{d-2}}{r^{d-2}} g$ is called the *kelvin transformation* relative to $S(a, \rho)$. The factor ρ^{d-2} ensures that $f = g$ on $S(a, \rho)$. This transformation is very useful in considerations of the Dirichlet problem for unbounded domains. The following result is an example of its application. See also Exercise 5.3.4.

Theorem 5.3.1. *A positive function u on the half-space $D = \{(x_1, \ldots, x_d) : x_d > 0\}$ is harmonic if and only if*

$$u(x) = Ax_d + x_d \int \frac{1}{|x - z|^d} m(dz) \qquad (5.3.1)$$

for a unique positive measure m on ∂D and a non-negative constant A.

Proof. It is simple to see that u defined by (5.3.1) is harmonic if it is finite at one point. Let $a = (0, 0, \ldots, 0, -2)$ and consider inversion relative to $\partial B(a, 2)$. Let $b = (0, 0, \ldots, 0, -1)$. From Figure 5.7, it is clear that the half-space D is mapped onto the interior of the ball $B(b, 1)$, with center b and radius 1.

The point a goes into the "point at infinity" while the rest of $\partial B(b, 1)$ is mapped onto ∂D. The Kelvin transformation gives us a positive harmonic function h in the ball $B(b, 1)$. From (5.2.5), there is a unique positive measure ν_1 on $S(b, 1)$, whose Poisson integral equals h. Letting $C = \nu_1(a)$

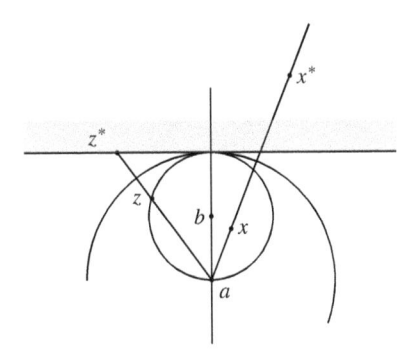

Fig. 5.7 Inversion transforming a half-space into the interior of a ball.

and $\nu = \nu_1 - C\delta_a$ (δ_a = point mass at a), we can write
$$h(x) = \frac{1 - |x - b|^2}{|x - a|^d}C + \left(1 - |x - b|^2\right)\int \frac{1}{|x - z|^d}\nu(dz).$$
The map $x \mapsto x^* = a + \frac{4}{|x-a|^2}(x - a)$ takes the measure ν into a measure μ on ∂D.

Since $(x^*)^* = x$ and $h(x) = \frac{4}{|x-a|^{d-2}}u(x^*)$, we get
$$u(x^*) = \frac{1 - |x - b|^2}{4|x - a|^2}C + \frac{1 - |x - b|^2}{4|x - a|^2}\int \frac{|x - a|^d}{|x - z^*|^d}\mu(dz)$$
or that
$$u(x^*) = \frac{1 - |x^* - b|^2}{4|x^* - a|^2}C + \frac{1 - |x^* - b|^2}{4|x^* - a|^2}\int \frac{|x^* - a|^d}{|x^* - z^*|^d}\mu(dz).$$
Using the definition of x^*,
$$|x^* - z^*|^2 = |(x^* - a) - (z^* - a)|^2$$
$$= \frac{16}{|x - a|^2} + \frac{16}{|z - a|^2} - \frac{32}{|x - a|^2|z - a|^2}\langle x - a, z - a\rangle$$
$$= \frac{16}{|x - a|^2|z - a|^2}|(x - a) - (z - a)|^2$$
$$= \frac{16}{|x - a|^2|z - a|^2}|x - z|^2.$$
And
$$|x^* - b|^2 = |x^* - a|^2 + |b - a|^2 - 2\langle x^* - a, b - a\rangle$$
$$= |x^* - a|^2 + 1 - \frac{8}{|x - a|^2}\langle x - a, b - a\rangle$$
$$= |x^* - a|^2 + 1 - \frac{1}{2}|x^* - a|^2(x_d + 2)$$
giving $1 - |x^* - b|^2 = \frac{1}{2}x_d|x^* - a|^2$. Thus,
$$u(x) = \frac{C}{8}x_d + \frac{1}{8}x_d\int \frac{1}{|x - z|^d}|z - a|^d\mu(dz)$$
which is of the form claimed. The uniqueness follows from Proposition 5.3.2. □

Proposition 5.3.2. *Let m be a signed measure on $\mathbb{R}^d$ such that*

$$\int \frac{1}{\left(1+|x|^2\right)^{\frac{d+1}{2}}} |m|\,(dx) < \infty, \tag{5.3.2}$$

where $|m|$ is the total variation measure of m. Then

$$v(x,t) = ct \int \frac{1}{\left(|x-y|^2 + t^2\right)^{\frac{d+1}{2}}} m\,(dx), \quad c = \frac{\Gamma\left(\frac{d+1}{2}\right)}{\pi^{\frac{d+1}{2}}} \tag{5.3.3}$$

is harmonic in the half-space $\{(x,t) : x \in \mathbb{R}^d,\, t > 0\}$ and m is the weak limit of $v(x,t)\,dx$, i.e.,

$$\lim_{t \to 0} \int f(x)\,v(x,t)\,dx = \int f(x)\,m(dx) \tag{5.3.4}$$

for every continuous function f with compact support.

Proof. Note that the Poisson integral $P(t,x,f)$ of f, i.e., the expression given by (5.2.6), converges to f boundedly, and for x far from the support of f, easy estimate on the denominator shows that $P(t,x,f)$ is comparable to $t\,|x|^{-d-1}$. The integral on the left side of (5.3.4) is, by Fubini, $\int m(dx)\,P(t,x,f)$. Due to (5.3.2) and dominated convergence, the limit in (5.3.4) can be taken inside. □

v is called the *Poisson integral of m* and denoted as $P(t,x,m)$.

Corollary 5.3.3. *Let v be bounded and harmonic in the half-space $\{(x,t) : x \in \mathbb{R}^d,\, t > 0\}$. Then there is a unique measure m satisfying (5.3.2) such that v is given by (5.3.3). In particular, if v is bounded and $\lim_{t \to 0} v(x,t) = 0$ for almost all x, then $v = 0$.*

Proof. Suppose $|v| \le 1$. Then $1 - v$ is positive and harmonic in the half-space, so by Theorem 5.3.1, there is a unique positive measure n such that

$$1 - v(x,t) = At + P(t,x,n).$$

Since $1 - v$ is bounded and n is positive, A must be zero. Also $1 = P(t,x,dy)$; dy denotes Lebesgue measure on $\mathbb{R}^d$. Since the Poisson integral of n is finite, n satisfies (5.3.3) so does of course dy. v is the Poisson integral of $dy - n$. The rest follows from Proposition 5.3.2. □

We leave it to the reader to derive the Poisson integral formula for a ball using the Kelvin transformation and the Poisson integral formula for a half-space (Exercise 5.2.7).

Exercises

5.3.1. Show that inversion takes spheres into spheres, hyperplanes being spheres of infinite radius. Specifically, show that inversion in $S(a,\rho)$ takes the sphere $S(Q,r)$ into the sphere with center $a + \rho^2 \frac{Q-a}{|Q-a|^2+r^2}$ and radius $\frac{r\rho^2}{||Q-a|^2-r^2|}$.

5.3.2. Show that two spheres one contained in the other can be transformed by inversion into concentric spheres. Specifically, let $S(a,r) \subset S(a,1)$. If $Q = \alpha a$, where α is a root of

$$\alpha^2 |a|^2 + \left(r^2 - |a|^2 - 1\right)\alpha + 1 = 0,$$

then inversion in $S(Q, \rho)$, $\rho > 0$ takes $S(a,r)$ and $S(0,1)$ into concentric spheres with center

$$Q - \rho^2 \frac{Q}{|Q|^2 - 1}.$$

5.3.3. Show that inversion takes a line into a line if and only if the line passes through the center of inversion.

5.3.4. Let $D = \{x : |x| > 1\}$. If u is harmonic in D, continuous on $\overline{D}$, and $\lim_{x \to \infty} u(x) = a$, then

$$u(x) = \int \frac{|x|^2 - 1}{|x-z|^d} u(z) \sigma(dz) + a\left[1 - \frac{1}{|x|^{d-2}}\right], \quad |x| > 1,$$

where $\sigma =$ uniform distribution on $S(0,1)$.

Hint: Use Kelvin transformation relative to $S(0,1)$ to get a function h, which is harmonic in the unit ball punctured at 0, satisfying $\lim_{x \to 0} |x|^{d-2} h(x) = a$. For each $0 < r < 1$, h is harmonic in the shell $\{x : r < |x| < 1\}$ and continuous on its closure. Now use (5.1.7), (5.1.8), and the Poisson integral formula of Example 5.2.6 to see that

$$h(x) = a\left[|x|^{-d+2} - 1\right] + \int \frac{1 - |x|^2}{|x-z|^d} h(z) \sigma(dz), \quad 0 < |x| < 1.$$

Transform back to arrive at the formula for u.

5.4 Boundary Limit Theorems of Fatou

Let m be a positive measure on $\mathbb{R}^d$ satisfying (5.3.2). Then $P(t, x, m)$, the Poisson integral of m, defined by (5.3.3) is harmonic in the upper half-space $\{(x,t) : x \in \mathbb{R}^d,\ t > 0\}$.

Define the *upper* and *lower* derivatives of m at x by

$$\overline{D}m(x) = \limsup_{r \to 0} \frac{m(B(x,r))}{|B(x,r)|},$$

$$\underline{D}m(x) = \liminf_{r \to 0} \frac{m(B(x,r))}{|B(x,r)|},$$

where $|B(x,r)|$ denotes the volume of the ball $B(x,r)$ with center x and radius r. The common value of the upper and lower derivatives (when they are equal) is denoted by $Dm(x)$. It is known that $Dm(x)$ exists for almost all x, relative to Lebesgue measure on $\mathbb{R}^d$. For a proof see Theorem 8.6 of Rudin [13, p. 154]. The following result is known as the "radial limit theorem" of Fatou.

Theorem 5.4.1. *With the above notation and $u(x,t) = P(t,x,m)$,*

$$\underline{D}m(x) \leq \liminf_{t \to 0} u(x,t) \leq \limsup_{t \to 0} u(x,t) \leq \overline{D}m(x). \tag{5.4.1}$$

In particular, for almost all x, $\lim_{t \to 0} u(t,x)$ exists and equals $Dm(x)$.

Proof. Let $x \in \mathbb{R}^d$ be fixed. If $F(r) = m(B(x,r))$, we can write

$$u(x,t) = c \int_{[0,\infty)} \frac{t}{(r^2 + t^2)^{\frac{d+1}{2}}} dF(r)$$

which when integrated by parts yields

$$u(x,t) = (d+1)c \int_{[0,\infty)} \frac{tr}{(r^2 + t^2)^{\frac{d+3}{2}}} F(r)\, dr$$

$$= (d+1)c \int_0^s \frac{F(r)}{r^d} \cdot \frac{tr^{d+1}}{(r^2+t^2)^{\frac{d+3}{2}}} dr$$

$$+ (d+1)c \int_s^\infty \frac{tr}{(r^2+t^2)^{\frac{d+3}{2}}} F(r)\, dr \tag{5.4.2}$$

for an arbitrary but fixed $s > 0$. The last integral tends to zero as t tends to zero because $\int_s^\infty r^{-d-1} F(r) \, dr$ is finite as is seen from $u(x,1) < \infty$. And

$$\lim_{r \to 0} c(d+1) \int_0^s \frac{tr^{d+1}}{(r^2+t^2)^{\frac{d+3}{2}}} dr = c(d+1) \int_0^\infty \frac{r^{d+1}}{(r^2+t^2)^{\frac{d+3}{2}}} dr = \frac{d}{\sigma},$$

where $\sigma =$ surface area of the unit sphere in $\mathbb{R}^d$. Thus, from (5.4.2) for all $s > 0$,

$$\limsup_{t \to 0} u(x,t) \leq \sup_{0 < r \leq s} F(r) \cdot \frac{d}{\sigma r^d}.$$

Since $|B(x,r)| = \frac{\sigma r^d}{d}$, we have proved the last part of (5.4.1). The first part is similar. □

We can strengthen the above theorem to include non-tangential limits: Let $K > 0$ be given. If $(x,t) \to (z,0)$ subject only to the condition that $|x-z| \leq Kt$, then for almost all z, $\lim u(x,t)$ exists and equals $Dm(z)$. In the figure, one sees the restriction imposed on the mode of convergence of (x,t) to $(z,0)$ and the reason is now clear for the name "non-tangential limit".

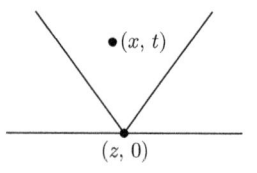

To prove the non-tangential limit result, let $m = m_a + m_s$, where m_a and m_s are the absolutely continuous and the singular parts of m. If Pm is the Poisson integral of m, $Pm = Pm_a + Pm_s$. Consider first Pm_s. Since

$$\left(|z-y|^2 + t^2\right)^{\frac{1}{2}} \leq \left(|x-y|^2 + t^2\right)^{\frac{1}{2}} + |x-z|,$$

if $|x-z| \leq Kt$,

$$\left(|z-y|^2 + t\right)^{\frac{d+1}{2}} \left(|x-y|^2 + t^2\right)^{-\frac{d+1}{2}}$$

$$\leq \left[1 + \frac{|x-z|}{\left(|x-y|^2 + t^2\right)^{\frac{1}{2}}}\right]^{d+1}$$

$$\leq \left[1 + \frac{|x-z|}{t}\right]^{d+1} (1+K)^{d+1} = M \quad \text{say}$$

and we get $(Pm_s)(x,t) \leq M(Pm_s)(z,t)$. $Dm_s(z) = 0$ for almost all z. Therefore, (5.4.1) implies that the non-tangential limit $(Pm_s)(x,t)$ exists and equals zero for almost all z.

Now consider Pm_a. If f denotes the density of m_a, f is locally integrable. It is known that (see Theorem 8.8 of Rudin [13, p. 158]) for almost all z,

$$\lim_{r \to 0} \frac{1}{|B(z,r)|} \int_{B(z,r)} |f(y) - f(z)| \, dy = 0. \tag{5.4.3}$$

Points for which (5.4.3) holds are called *Lebesgue points* for f. If z_0 is a Lebesgue point for f and we define the measure a by

$$a(dy) = |f(y) - f(z_0)| \, dy,$$

$Da(z_0)$ exists and equals 0. As in the case of Pm_s, we have

$$(Pa)(x,t) \leq M(Pa)(z,t)$$

for any (x,t), (z,t) satisfying $|x - z| \leq Kt$. One concludes from (5.4.1) that

$$\lim (Pa)(x,t) = 0, \text{ as } (x,t) \to (z_0, 0) \text{ and } |x - z_0| \leq Kt,$$

i.e., that $\lim (Pm_a)(x,t)$ exists and equals $f(z_0)$ if (x,t) tends to $(z_0, 0)$ subject only to the condition $|x - z_0| \leq Kt$. This proves the following:

Theorem 5.4.2 (Non-tangential Limit Theorem of Fatou). *Let u be the Poisson integral of m. For almost all z, the following statement holds: If (x,t) tends to $(z, 0)$ subject only to the condition $|x - z| \leq Kt$ where K is some positive number (which may depend on z), $\lim u(x,t)$ exists and equals $Dm(z)$.*

The above non-tangential result together with a Kelvin transformation can be used to get corresponding non-tangential limit theorems for harmonic functions in a ball. We leave the details to the reader.

An application. Let $p \geq 1$ and $f \in L^p(-\infty, \infty)$; for each $\epsilon > 0$, let

$$H_\epsilon(x, f) = \frac{1}{\pi} \int_{|x-t| \geq \epsilon} \frac{f(t)}{x - t} \, dt. \tag{5.4.4}$$

We show below that

$$H(x, f) = \lim_{\epsilon \to 0} H_\epsilon(x, f) \tag{5.4.5}$$

exists almost everywhere. $H(x, f)$ is called the *Hilbert transform* of f and written as

$$H(x, f) = \frac{1}{\pi} \int_{-\infty}^{\infty} \frac{f(t)}{x - t} \, dt.$$

For an account of the importance of Hilbert transform in Fourier series, refer to Chapter 4 of Garsia [10]. We may assume that $f \geq 0$. The Poisson integral

$$u(x,y) = \frac{1}{\pi}\int_{-\infty}^{\infty}\frac{f(t)\cdot y}{(x-t)^2+y^2}dt \qquad (5.4.6)$$

is harmonic in the upper half plane ($y > 0$) and tends as $y \to 0$ to $f(x)$ at every Lebesgue point x of f. The function

$$v(x,y) = \frac{1}{\pi}\int_{-\infty}^{\infty}f(t)\frac{x-t}{(x-t)^2+y^2}dt \qquad (5.4.7)$$

is called the *conjugate Poisson integral* of f; the nomenclature is justified because $u + iv$ is the analytic (in the half plane) function

$$\frac{i}{\pi}\int_{-\infty}^{\infty}\frac{f(t)}{z-t}dt, \quad z = x+iy,\ y > 0.$$

The conjugate Poisson kernel $\frac{1}{\pi}\frac{x-t}{(x-t)^2+y^2}$ is not integrable. However, it is bounded and is in L^q for all $q > 1$. The integral in (5.4.7) thus makes sense for all $f \in L^p$, $p \geq 1$. Since

$$\int_{|x-t|>y}\frac{1}{(x-t)\left((x-t)^2+y^2\right)}dt = \int_{|x-t|\leq y}\frac{(x-t)}{(x-t)^2+y^2}dt = 0,$$

we easily see that

$$|H_y(x,f) - v(x,y)| \leq \frac{1}{\pi}\int_{|x-t|>y}|f(t)-f(x)|\frac{y}{|x-t|\left((x-t)^2+y^2\right)}dt$$

$$+ \int_{|x-t|\leq y}|f(t)-f(x)|\frac{|x-t|}{(x-t)^2+y^2}dt$$

$$\leq \int_{-\infty}^{\infty}|f(t)-f(x)|\frac{y}{(x-t)^2+y^2}dt.$$

We have shown that the right side tends as $y \to 0$ to zero provided x is a Lebesgue point for f. Thus, for almost all x, $|H_y(x,f) - v(x,y)| \to 0$ as $y \to 0$. Hence, we need only show that $\lim_{y \to 0} v(x,y)$ exists for almost all x. The function $e^{-(u+iv)}$ is analytic in the upper half-plane and bounded by 1 since $u \geq 0$. It follows that $\lim_{y \to 0} e^{-(u+iv)}$ exists almost everywhere (the real and imaginary parts of $e^{-(u+iv)}$ are bounded harmonic functions in the upper half plane, see Exercise 5.4.1). Since $\lim_{y \to 0} u(x,y)$ exists and is finite almost everywhere, one deduces that $\lim_{y \to 0} e^{-iv(x,y)}$ exists almost everywhere. Since for each x, $v(x,y)$ is continuous in y, $\lim_{y\to 0} v(x,y)$ exists almost everywhere (Exercise 5.4.3). This establishes the existence of the Hilbert transform. If $p > 1$, the Hilbert transform H maps L^p onto L^p, is continuous, and $H^2 = -I$ (I = identity), and for $p = 2$, it is an isometry. These non-trivial results follow from maximal inequalities. For real variables proofs, see Chapter 4 of Garsia [10].

Exercises

5.4.1. Let $u(x,t)$ be bounded and harmonic in the half-space $\{(x,t) : x \in \mathbb{R}^d, t > 0\}$. Then, $g(x) = \lim_{t\to 0} u(x,t)$ exists almost everywhere and u is the Poisson integral of g.

Hint: By Corollary 5.3.3, u is the Poisson integral of a measure m. By 5.4.1, $g(x) = \lim_{t\to 0} u(x,t)$ exists almost everywhere. Now use the last part of Proposition 5.3.2 and bounded convergence.

5.4.2. Let $u(x,t)$ be harmonic in the half-space. u is the Poisson integral of a function $f \in L^p$, $1 < p < \infty$, if and only if $\sup_t \|u(\cdot, t)\|_p < \infty$, $\|\cdot\|_p$ denoting L^p-norm. u is the Poisson integral of a finite-signed measure if and only if $\sup_t \|u(\cdot, t)\|_1 < \infty$.

Hint: By the mean value property, $|u(x,t)|^p \leq$ the volume average of $|u(y,s)|^p$ on the ball of radius t and center (x,t). Therefore (see figure),

$$|u(x,t)|^p \leq \text{const} \int_0^{2t} \int_{\mathbb{R}^d} |u(y,s)|^p \, dy \leq \text{const} \cdot t^{-d}.$$

Hence, for each $t > 0$, $u(x, t+s)$ is bounded and harmonic on the half-space $\{(x, s) : s > 0\}$. By Exercise 5.4.1,

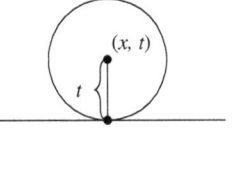

$$u(x, t+s) = c \int \frac{s}{\left(|x-y|^2 + s^2\right)^{\frac{d+1}{2}}} u(y, s) \, dy.$$

If $\sup_s \|u(\cdot, s)\|_1 < \infty$, let m be a weak limit of the measures $u(x, s) \, dx$ (as s tends to zero). If $p > 1$ and $\sup_s \|u(\cdot, s)\|_p < \infty$, let f be a weak limit of $u(\cdot, s)$ as s tends to zero. Note that the Poisson kernel is continuous, vanishes at ∞, and $\in L^q$ for all $q \geq 1$.

5.4.3. Let $-\infty \leq \lambda < \mu \leq \infty$ and f be real and continuous on the open interval (λ, μ). If $\lim_{x\to\lambda} e^{if(x)}$ exists, so does $\lim_{x\to\lambda} f(x)$ as a finite limit.

5.5 Spherical Harmonics

To a function on the circle corresponds its Fourier series. The functions $e^{in\theta}$ are the restrictions to the unit circle of z^n whose real and imaginary parts are homogeneous harmonic polynomials. Any homogeneous polynomial of degree n in $x_1, \ldots, x_d$ is called an nth *solid harmonic* of order n in d-dimensions. Its restriction to the unit sphere is called an nth *surface or spherical harmonic of order n*. For example, a constant is a solid harmonic of order 0, $\sum a_i x_i$ (a_i constant) is the most general solid harmonic of order 1, $\sum a_{ij} x_i x_j$ with $\sum a_{ii} = 0$ is the most general solid harmonic or order 2, and so on. In this section, we briefly look at some properties of surface harmonics and expand a function in terms of surface harmonics analogous to Fourier expansion on the circle.

Our starting is the Poisson kernel for the unit ball:

$$P(x,z) = \left(1 - |x|^2\right)|x-z|^{-d}, \quad |x| < 1, \ |z| = 1. \tag{5.5.1}$$

Writing $\mu = \frac{x}{|x|} \cdot z =$ inner product of $\frac{x}{|x|}$ and z, and $r = |x|$, $P(x,z)$ takes the form

$$\left(1-r^2\right)\left[1 - 2\mu r + r^2\right]^{-d/2}. \tag{5.5.2}$$

If $\left|2\mu r - r^2\right| < 1$, we can expand the function in (5.5.2) as a power series in $(2\mu r - r^2)$. It is however more useful to write this a power series in r. This involves rearrangement of terms, which is justifiable if the series is absolutely convergent. Absolutely convergence is clearly guaranteed if $2|\mu r| + |r^2| < 1$ which ($|\mu| \le 1$) is valid if $|r| < \sqrt{2} - 1$.

Thus, if $|r| < \sqrt{2} - 1$, we can expand the function in (5.5.2) as a power series in $r(2\mu - r)$ and rearrange the series as a power series in r. It is clear that the coefficient of r^n is a *polynomial in μ of degree exactly n*:

$$\frac{1-r^2}{(1-2\mu r + r^2)^{d/2}} = \sum_0^\infty P_n(\mu) r^n, \quad |r| < \sqrt{2}-1, \ |\mu| \le 1. \tag{5.5.3}$$

P_n are called Gegenbauer or hyperspherical polynomials.

Let us emphasize that the series in (5.5.3) converges uniformly in μ for $|\mu| \le 1$ if we replace each coefficient of the polynomial P_n by its absolutely value. This fact easily justifies term by term differentiation relative to μ.

Thus, writing $\mu = \frac{x}{|x|} \cdot z$,

$$\frac{1-|x|^2}{|x-z|^d} = \sum |x|^n P_n(\mu), \quad |x| < \frac{1}{3}, \ |z| = 1. \tag{5.5.4}$$

We soon show that the above is valid for $|x| < 1$. For fixed z, $|z| = 1$, the left side in (5.5.4) is harmonic for $|x| < 1$. Since P_n is a polynomial, $|x|^n P_n(\mu)$ is infinitely differentiable except perhaps at $x = 0$. Term by term differentiation is justified:

$$\Delta(|x|^n P_n(\mu))$$
$$= |x|^{n-2}\{(1-\mu^2) P_n''(\mu) - (d-1)\mu P_n'(\mu) + n(n+d+2) P_n(\mu)\},$$

except perhaps at $x = 0$. We get

$$\sum_0^\infty \Delta(|x|^n P_n(\mu)) = 0, \quad |x| < \frac{1}{3}, \ x \neq 0.$$

$|x|^n P_n(\mu)$ is homogeneous of degree n and $\Delta(|x|^n P_n(\mu))$ is homogeneous of degree $n-2$ ($n \geq 2$; $P_0(\mu) = 1$, $|x| P_1(\mu)$ is easily seen to be a polynomial of degree 1 and hence harmonic). Such a sum cannot be identically zero unless each term is zero (because for $\frac{x}{|x|}$ and z fixed, it becomes a power series in $r = |x|$). Thus, for each n, $|x|^n P_n(\mu)$ is harmonic for $|x| < \frac{1}{3}$ except perhaps at 0; since it is continuous at 0, it must be harmonic at 0 as well. Since it is homogeneous of order n, it must be a homogeneous polynomial of degree n. See Exercises 5.5.1 and 5.5.2. Thus, we have shown the following:

A. *For each $z \in \partial B(0,1)$, $|x|^n P_n\left(\frac{x}{|x|} \cdot z\right)$ is a solid harmonic of order n, and for each $z \in \partial B(0,1)$, $P_n(x,z)$ (($|x| = 1$)) is a spherical harmonic of order n.*

A clearly implies the following (since the space of solid harmonics of order n is finite dimensional):

B. *For each finite measure μ on $\partial B(0,1)$,*

$$|x|^n \int P_n\left(\frac{x}{|x|} \cdot z\right) \mu(dz)$$

is a solid harmonic of order n and

$$\int P_n(\xi \cdot z) \mu(dz), \quad |\xi| = 1,$$

a spherical harmonic of order n.

If u is continuous on the closed unit ball and harmonic inside,

$$u(x) = \int_{|x|=1} \frac{1-|x|^2}{|x-z|^d} u(z) p(dz)$$

p being the uniform distribution on $|x|=1$. Using (5.5.4),

$$u(x) = \sum_0^\infty |x|^n \int P_n\left(\frac{x}{|x|} \cdot z\right) u(z) p(dz), \quad |x| < \frac{1}{3}, \qquad (5.5.5)$$

the nth term is a solid harmonic of order n. In particular, if u is a solid harmonic of order n, Equation (5.5.5) expresses a homogeneous function of order n as a sum, for $|x| < \frac{1}{3}$, of homogeneous functions of order 0, 1, 2, etc. We conclude that all terms except the nth must vanish. We have thus shown the following:

C. *If H is a solid harmonic of order n, then*

$$H(x) = |x|^n \int_{|x|=1} P_n\left(\frac{x}{|x|} \cdot z\right) H(z) p(dz),$$

$$0 = \int_{|z|=1} P_m\left(\frac{x}{|x|} \cdot z\right) H(z) p(dz), \quad m \neq n. \qquad (5.5.6)$$

The second identity in (5.5.6) is identical to saying that a spherical harmonic of order n is orthogonal (relative to p) to $P_m(\xi \cdot z)$ (regarded as a function of z) for each $|\xi| = 1$ and $m \neq n$, while the first identity says that if S_n is a surface harmonic of order n,

$$S_n(x) = \int_{|z|=1} P_n(x \cdot z) S_n(z) p(dz), \quad |x| = 1. \qquad (5.5.7)$$

The set of all solid harmonics of order n is finite dimensional. The first formula in (5.5.6) shows that the solid harmonics

$$|x|^n P_n\left(\frac{x}{|x|} \cdot z\right)$$

span this space as z varies on $\partial B(0,1)$. Thus, we have shown the following:

D. *There is a finite subset $F \subset \partial B(0,1)$ such that every surface harmonic S_n of order n can be written as*

$$S_n(x) = \sum_{z \in F} c_z P_n(x \cdot z), \quad |x| = 1,$$

where c_z are constants.

We have shown in **C** that a surface harmonic of order n is orthogonal to $P_m(x,z)$ for each $|x| = 1$ and $m \neq n$. This together with **D** gives us the following:

E. *Two surface harmonics of distinct degrees are orthogonal (with respect to p).*

Let $S_{n,j}$ be a total orthonormal set of nth spherical harmonics (i.e., $S_{n,j}$ are orthonormal relative to p and span the space of all n-spherical harmonics). For each ξ, $|\xi| = 1$, $P_n(\xi \cdot \eta)$ is as a function of η a spherical harmonic of order n. We can therefore write

$$P_n(\xi \cdot \eta) = \sum_j a_j(\xi) S_{n,j}(\eta).$$

The coefficients $a_j(\xi)$ are given by

$$a_j(\xi) = \int P_n(\xi \cdot \eta) S_{n,j}(\eta) p(d\eta).$$

Equation (5.5.7) implies that $a_j(\xi) = S_{n,j}(\xi)$. Thus, we have proved the following:

F. *(Addition theorem) Let $S_{n,j}$ be a total orthonormal set of spherical harmonics of order n. Then,*

$$\sum_j S_{n,j}(\xi) S_{n,j}(\eta) = P_n(\xi \cdot \eta), \quad |\xi| = |\eta| = 1. \tag{5.5.8}$$

Putting $\xi = \eta$ in (5.5.8), we get

$$\sum_j (S_{n,j}(\xi))^2 = P_n(1). \tag{5.5.9}$$

Integrating with respect to σ and remembering that $\int (S_{n,j}(\xi))^2 \sigma(d\xi) = 1$, we get the following:

G. $P_n(1) =$ *dimension of the space of spherical harmonics of order n.*

$P_n(1)$ can easily be calculated as follows. Writing

$$(1 - 2\mu r + r^2)^{-\frac{d}{2}} = \sum_0^\infty Q_n(\mu) r^n,$$

we get

$$(1 - r^2)(1 - 2\mu r + r^2)^{-\frac{d}{2}} = Q_0(\mu) + rQ_1(\mu) + \sum_2^\infty r^n (Q_n(\mu) - Q_{n-2}(\mu))$$

so that $P_n(\mu) = Q_n(\mu) - Q_{n-2}(\mu)$ for $n \geq 2$. $Q_n(1)$ is the coefficient of r^n in $(1 - 2\mu r + r^2)^{-\frac{d}{2}}$, i.e.,

$$(1-r)^{-d} = \sum Q_n(1) r^n,$$

giving $Q_1(1) = \binom{n+d-1}{n}$ and therefore

$$P_n(1) = \binom{n+d-1}{n} - \binom{n+d-3}{n-2}, \quad n \geq 2,$$

$$P_1(1) = d,$$

$$P_0(1) = 1.$$

Thus, we have shown the following:

H. *The space of spherical harmonics of order n has dimension*

$$\binom{n+d-1}{n} - \binom{n+d-3}{n-2}, \quad n \geq 2.$$

Equations (5.5.8) and (5.5.9) imply, using the Cauchy–Schwarz inequality,

$$|P_n(\xi \cdot \eta)| \leq \sum_j |S_{n,j}(\xi) S_{n,j}(\eta)|$$

$$\leq \left(\sum_j (S_{n,j}(\xi))^2 \right)^{1/2} \left(\sum_j (S_{n,j}(\eta))^2 \right)^{1/2} = P_n(1)$$

proving that

$$|P_n(t)| \leq P_n(1), \quad -1 \leq t \leq 1. \tag{5.5.10}$$

Combining (5.5.10) and the value of $P_n(1)$ found in **G**, we see that the series in (5.5.4) is uniformly convergent on compact subsets of the open ball $B(0,1)$. It therefore represents a harmonic function in $B(0,1)$. Thus, (5.5.4) is valid for $|x| < 1$ (harmonic functions are analytic!).

One easily deduces from the above that if u is harmonic in the ball $B(0,R)$ and continuous on its closure, then for $|x| < R$,

$$u(x) = \sum_0^\infty \frac{|x|^n}{R^n} \sum_1^{P_n(1)} S_{n,j}\left(\frac{x}{|x|}\right) \int_{|z|=1} S_{n,j}(z) u(Rz) p(dz) \tag{5.5.11}$$

(use (5.5.8) and **G**). This is the expansion of u in harmonics analogous to the Fourier expansion in two dimensions.

As a simple application, we have the following:

Theorem 5.5.1 (Liouville's Theorem). *Let u be harmonic on $\mathbb{R}^d$ and suppose that for some N, $u(x) \leq |x|^N$ for all large $|x|$. Then u is a polynomial of degree at most N.*

Proof. In (5.5.11), the coefficient

$$a_{n,j} = \frac{1}{R^n} \int_{|z|=1} S_{n,j}(z) u(Rz) p(dz)$$

does not depend on R. Indeed, writing rx ($r < R$) instead of x in (5.5.11), with $|x| = 1$, multiplying by $S_{n,j}$ and integrating, we get

$$\int_{|x|=1} u(rx) S_{n,j}(x) p(dx) = \frac{r^n}{R^n} \int_{|z|=1} S_{n,j}(z) u(Rx) p(dx)$$

as asserted. Let $n > N$ and $b_{n,j} = \sup_{|x|=1} |S_{n,j}(x)|$. We get

$$\int_{|x|=1} (b_{n,j} \pm S_{n,j}(x)) u(rx) p(dx) = b_{n,j} u(0) \pm a_{n,j} r^n.$$

Since $b_{n,j} \pm S_{n,j}(x) \geq 0$ and $u(rx) \leq r^N$,

$$b_{n,j} u(0) \pm a_{n,j} r^n \leq b_{n,j} r^N$$

since $\int S_{n,j}(x) p(dx) = 0$ (because $\int_{|x|=1} S_{n,j}(x) \sigma(dx) =$ the value at 0 of a homogeneous polynomial of degree n). This inequality for all large r implies $a_n = 0$ for all $n > N$. □

That P_n has degree n implies an interesting result on harmonic functions with polynomial values on $\partial B(0,1)$. Since P_n has degree n for each n, we can write t^n as a linear combination of polynomials P_k, $0 \leq k \leq n$, i.e., we can write

$$t^n = \sum_0^n a_k P_k(t)$$

implies

$$(x \cdot y)^n = \sum_{k \leq n} a_k P_k(x \cdot y), \quad x, y \in \mathbb{R}^d. \tag{5.5.12}$$

For $|x| = 1$, $P_k(x \cdot y)$ (as a function of x on $S(0,1)$) is a spherical harmonic of order k for each y with $|y| = 1$. Equation (5.5.12) says that for each y

with $|y| = 1$ (and hence for any $y \in \mathbb{R}^d$), the homogeneous polynomial $Q(x) = (x \cdot y)^n$ agrees with a harmonic polynomial of degree at most n on the surface of the unit ball. It can be shown (see Exercise 5.5.3) that every homogeneous polynomial of degree n is a linear combination of polynomials of the form $Q(x) = (x \cdot y)^n$ as y ranges over $\mathbb{R}^d$. We have thus shown the following:

I. *To every polynomial of degree n corresponds a unique harmonic polynomial of degree at most n which agrees with the given polynomial on $S(0,1)$.*

See Exercise 5.5.4 for the so-called Funk–Hecke formula. In the exercises, some properties of P_n as well as an application are given.

Example 5.5.2. When $d = 2$ from **G** there are only 2 linearly independent solid harmonics of degree n, namely, the real and imaginary parts of z^n.

Exercises

5.5.1. If u is bounded and harmonic in the ball $B(0,1)$ punctured at 0, it can be defined to be harmonic at 0 as well.
Hint: Let $r < 1$ be fixed. For $s < |x| < r$,
$$u(x) = \mathbb{E}_x\left[u(X_{T_r}) : T_r < T_s\right] + \mathbb{E}_x\left[u(X_{T_s}) : T_s < T_r\right],$$
where T_s and T_r are the hitting times to $B(0,s)$ and $\partial B(0,r)$, respectively. As s tends to zero, the second term tends to zero.

5.5.2. Let u be homogeneous of degree n. If u is C^n in a neighborhood of 0, then u is a polynomial of degree n.
Hint: $\frac{\partial u}{\partial x_i}$ is homogeneous of degree $n-1$ and is C^{n-1} in a neighborhood of 0. Use induction.

5.5.3. (Using induction) show that every homogeneous polynomial of degree m in $(x_1, \ldots, x_d)$ is a linear combination of polynomials of the form $(\xi \cdot \eta)^m$ as ξ varies in $\mathbb{R}^d$.
Hint: Induction will be on the number of variables. So assume the result is true for all homogeneous polynomials of arbitrary degree in $d-1$ variables. It is clearly sufficient to show the result for any polynomial of the form $x_1^{i_1} \cdots x_{d-1}^{i_{d-1}}$ with $i_1 + \cdots + i_d = m$. By induction assumption, $x_1 \cdots x_{d-1}$ is a linear combination of expressions of the form $(x \cdot \xi)^{m-i_d}$,

ξ ranging over $\mathbb{R}^{d-1}$ or ξ ranging over $\mathbb{R}^d$ with $\xi_d = 0$. Thus, it suffices to show the result for each polynomial of the form $(x \cdot \xi)^i x_d^j$, $i + j = m$, $\xi_d = 0$. If $\eta_k = \xi_k$, $k \le d - 1$, and $\eta_d = \alpha_k$, $(x \cdot \eta) = (x \cdot \xi) + \alpha x_d$. We have

$$((x \cdot \xi) + \alpha_k x_d)^m = \sum \binom{m}{i} (x \cdot \xi)^{m-i} \alpha_k^i x_d^i.$$

If the α's are distinct, the Vandermonde determinant

$$\begin{vmatrix} 1 & \cdots & \alpha_1 & \cdots & \alpha_1^m \\ \vdots & & & & \\ 1 & \cdots & \alpha_{m+1} & \cdots & \alpha_{m+1}^m \end{vmatrix}$$

is not zero (see Bourbaki [5, p. 99], Chapter III), and therefore each $\binom{m}{i} (x \cdot \xi)^{m-i} x_d^i$ is a linear combination of $((x \cdot \xi) + \alpha x_d)^m = (x \cdot \eta)^m$.

5.5.4 (Funk–Hecke Formula). Let S_n be a spherical harmonic of order n and f continuous on $[-1, 1]$. Show that

$$\int f(\xi \cdot \eta) S_n(\eta) p(d\eta) = \lambda S_n(\xi)$$

for some constant λ.

Hint: Let m be the measure on $[-1, 1]$ induced by the map $\eta \mapsto \xi \cdot \eta$, $\eta \in S(0, 1)$. From **C**, $\int_{-1}^{1} P_n(t) P_m(t) m(dt) = 0$ if $n \ne m$. Since t^n is a linear combination of P_k, $k \le n$, $\{P_k(t)\}$ is total in $L^2(dm)$. If $f \in L^2(dm)$,

$$f = \sum a_n P_n \quad (L^2\text{-sense}),$$

i.e., $f(\xi \cdot \eta) = \sum a_n P_n(\xi \cdot \eta)$. Now use (5.5.7) and **E**.

5.5.5. Show that P_n satisfy

$$(1 - t^2) P_n''(t) + (1 - d) t P_n'(t) + n(n + d - 2) P_n(t) = 0.$$

Hint: Use that $|x|^n P_n\left(\frac{x}{|x|} \cdot y\right)$ is harmonic.

Remark. The differential equation satisfied by P_n may be written as

$$\frac{d}{dt}\left[(1-t^2)^{\frac{d-1}{2}} P_n'\right] + n(n+d-2)(1-t^2)^{\frac{d-3}{2}} P_n = 0.$$

This has the familiar Sturm–Liouville form and it is general knowledge (one can verify this directly, see, for instance, Birkhoff [2], Chapter X) that P_n

must be orthogonal on $[-1,1]$ relative to the weight function $(1-t^2)^{\frac{d-3}{2}}$. If m is the probability measure given by the map $y \mapsto x \cdot y$ of $S(0,1)$ onto $[-1,1]$, we also know that P_n are orthogonal relative to m. This implies m has density $c(1-t^2)^{\frac{d-3}{2}}$, c being chosen to make the integral 1; indeed, one sees inductively that

$$\int_{-1}^1 t^n dm = c \int_{-1}^1 (1-t^2)^{\frac{d-3}{2}} t^n dt.$$

5.5.6. Show that every function u which is harmonic in a shell $\{x : r < |x| < R\}$ can be written as $u = v + w$, where v is harmonic in $\{x : |x| < R\}$ and w is harmonic in $\{x : |x| > r\}$.

Hint: Using Kelvin transformation, we see that $|x|^{-n-d+2} S_{i,n}\left(\frac{x}{|x|}\right)$ is harmonic except at the origin, where $S_{i,n}$ is a spherical harmonic of order n. For the exercise, it is clearly enough to assume that u is continuous in the closed shell (otherwise just look at smaller shells) with boundary values $f(x)$ on $|x| = r$ and $g(x)$ on $|x| = R$. Consider formally the series

$$\sum_{n=1}^\infty \sum_i \left(a_{i,n}|x|^n b_{i,n}|x|^{-n-d+2}\right) S_{i,n}\left(\frac{x}{|x|}\right)$$

where for each n, $S_{i,n}$ form a complete set of orthonormal spherical harmonics of degree n. Compute the "Fourier coefficients" $a_{i,n}$ and $b_{i,n}$ by

$$a_{i,n} r^n + b_{i,n} r^{-n-d+2} = \int_{|x|=1} f(rx) S_{i,n}(x) p(dx),$$

$$a_{i,n} R^n + b_{i,n} R^{-n-d+2} = \int_{|x|=1} g(Rx) S_{i,n}(x) p(dx),$$

where p is the uniform distribution on the unit sphere $\{x : |x| = 1\}$. It is simple to show that the series

$$v(x) = \sum_{n \geq 1} |x|^n \sum_i a_{i,n} S_{i,n}\left(\frac{x}{|x|}\right)$$

converges uniformly on compact subsets of $\{x : |x| < R\}$, and

$$h(x) = \sum_{n \geq 1} |x|^{-n-d+2} \sum_i b_{i,n} S_{i,n}\left(\frac{x}{|x|}\right)$$

converges uniformly on compact subsets of $\{x : |x| > r\}$. Thus, $v(x) + h(x)$ is harmonic in $\{x : r < |x| < R\}$. Since $S_{i,n}$, $n = 0, 1, 2, \ldots$, form a complete set in $L^2(dp)$, the difference $u-h$ can only assume constant values on $|x|=r$ and $|x|=R$ and the most general such is

$$a + b\log|x|, \quad d = 2,$$
$$a + b|x|^{-d+2}, \quad d \geq 3.$$

Thus, u has the expansion

$$u(x) = a + b\log|x| + \sum_{n \geq 1}\left(\sum_i a_{i,n}|x|^n + b_{i,n}|x|^{-n}\right) S_{i,n}\left(\frac{x}{|x|}\right), \quad d = 2,$$

$$u(x) = \sum_{n=0}^{\infty}\left(\sum_i a_{i,n}|x|^n + b_{i,n}|x|^{-n-d+2}\right) S_{i,n}\left(\frac{x}{|x|}\right), \quad d \geq 3.$$

For more on spherical harmonics, consult MacRobert [12]. For applications to representation theory, start with Chapter 4 of Dym and McKean [7].

References

[1] S. Axler, P. Bourdon and W. Ramey. *Harmonic Function Theory*. Graduate Texts in Mathematics, Vol. 137. Springer-Verlag, New York, 1992.

[2] G. Birkhoff and G.-C. Rota. *Ordinary Differential Equations*, 2nd edn. Blaisdell Publishing Co. [Ginn and Co.], Waltham, 1969.

[3] A. Björn, J. Björn and X. Li. Sphericalization and p-harmonic functions on unbounded domains in Ahlfors regular spaces. *J. Math. Anal. Appl.*, 474(2): 852–875, 2019.

[4] D. Borthwick. *Spectral Theory — Basic Concepts and Applications*. Graduate Texts in Mathematics, Vol. 284. Springer, Cham, 2020.

[5] N. Bourbaki. *Éléments de mathématique. Algèbre. Chapitres 1 à 3*. Hermann, Paris, 1970.

[6] X. Chen. Dynkin's formula under the G-expectation. *Stat. Probab. Lett.*, 80(5–6): 519–526, 2010.

[7] H. Dym and H. P. McKean. *Fourier Series and Integrals*. Probability and Mathematical Statistics, No. 14. Academic Press, New York, 1972.

[8] W. Feller. *An Introduction to Probability Theory and Its Applications*, 2nd edn., Vol. II. John Wiley & Sons, Inc., New York, 1971.

[9] J. B. Garnett and D. E. Marshall. *Harmonic Measure*. New Mathematical Monographs, Vol. 2. Cambridge University Press, Cambridge, 2005.

[10] A. M. Garsia. *Topics in Almost Everywhere Convergence*. Lectures in Advanced Mathematics, No. 4. Markham Publishing Co., Chicago, 1970.

[11] B. Jaye and F. Nazarov. Reflectionless measures for Calderón-Zygmund operators II: Wolff potentials and rectifiability. *J. Eur. Math. Soc. (JEMS)*, 21(2): 549–583, 2019.

[12] T. M. MacRobert. *Spherical Harmonics. An Elementary Treatise on Harmonic Functions with Applications*. International Series of Monographs in Pure and Applied Mathematics, Vol. 98. Pergamon Press, Oxford, 1967. Third edition revised with the assistance of I. N. Sneddon.

[13] W. Rudin. *Real and Complex Analysis*. McGraw-Hill Book Co., New York, 1966.

[14] N. B. Salem. Inequalities related to spherical harmonics associated with the Weinstein operator. *Integral Transforms Spec. Funct.*, 34(1): 41–64, 2023.

[15] A. V. Swishchuk. The analogue of Dynkin's formula and boundary value problems for multiplicative operator functionals of Markov processes. *Akad. Nauk Ukrainy Inst. Mat. Preprint*, (44): 16, 1993.

[16] Y. Wang. Strichartz's Radon transforms for mutually orthogonal affine planes and fractional integrals. *Fract. Calc. Appl. Anal.*, 25(5): 1971–1993, 2022.

Chapter 6

Superharmonic Functions and Riesz Measures

In this chapter, we extend the results from Chapter 5 to the important context of superharmonic functions.

A key result in probabilistic potential theory is the fact that superharmonic functions, i.e., typically quite irregular lower semicontinuous functions, probabilistically may be seen as continuous functions along Brownian paths. This discovery by Doob has opened an exciting communication between the following two areas: (i) continuous time martingale analysis and (ii) supermartingale theory and potential theory. And this is in a manner where each one serves to influence the other in fruitful ways.

The original material in Chapter 5 dealing with superharmonic functions and applications is still current, and we have added a discussion of new directions. In addition, for the benefit of readers, we offer the following citations covering new directions [1, 4, 6–11, 13].

The new directions include generalized sweeping-out and probability, Schrödinger–Green potential and boundary behavior of superharmonic functions, and convexity estimates for a class of semilinear elliptic problems.

For the benefit of beginner readers, we offer the following supplementary introductory text [2].

Introduction. Superharmonic functions can be regarded as generalizations of concave functions and bear the same relationship to harmonic functions as concave functions do to linear functions.

The theory of superharmonic functions was started by F. Riesz. For a little history, we refer the reader to Rado [15] and the references cited here. In Rado [15], there is also a discussion of a wide range of applications of superharmonic functions. We only give some applications. In any case, the reader sees in the sequel that the study of superharmonic functions more than justifies itself by its applications to the theory of harmonic functions alone.

In Section 6.1, we define and investigate some general properties of superharmonic functions. Section 6.2 is devoted to applications and Section 6.3 deals with Riesz measures associated with superharmonic functions. Section 6.4 is concerned with continuity properties of superharmonic functions, and finally, in Section 6.5, we show the uniqueness of the solution of the modified Dirichlet problem.

6.1 Superharmonic Functions

$B(a, r)$ denotes a ball with center a and radius r and $S(a, r) = \partial B(a, r)$ its topological boundary.

Let G be an open set in $\mathbb{R}^d$. A function $f : G \to (-\infty, \infty)$ is called *superharmonic* on G if

(S.1) f is lower semicontinuous on G,
(S.2) $f(a) < \infty$ at a dense set of points in G,
(S.3) for every a in G and every ball $B(a, r) \subset G$, $\mathbb{E}_a [f(X_T)] \leq f(a)$, where T is the exit time from $B(a, r)$.

f is called *sub-harmonic* if $-f$ is superharmonic.

If only (S.1) and (S.3) hold, we say that f is *hyperharmonic*. If G is a domain, the only function that is hyperharmonic and not superharmonic is the function identically equal to ∞. See Exercise 6.1.1 and the remark thereafter. It is easy to see that if f is hyperharmonic so is $f \wedge n$ for all real n. Thus, every hyperharmonic function is an increasing limit of a sequence of superharmonic functions that are bounded above.

Condition (S.3) can be rewritten as $\int_{S(a,r)} f(b) \, d\sigma(b) \leq f(a)$, where σ is the uniform distribution on $S(a, r)$. Integration using polar coordinates then implies

$$\frac{1}{P_r} \int_{B(a,r)} f(b) \, db \leq f(a), \quad P_r = \text{volume of } B(a, r). \tag{6.1.1}$$

By lower semi-continuity of f and (6.1.1),

$$\lim_{r \to 0} \frac{1}{P_r} \int_{B(a,r)} f(b) \, db = f(a). \tag{6.1.2}$$

Indeed, from (6.1.1), $\limsup_{r\to 0} \int_{B(a,r)} f(b) \, db \leq f(a)$; given a number $c < f(a)$, by lower semi-continuity, we can find r so that $\inf_{b \in B(a,r)} f(b) > c$ and this at once gives the above claim. Thus, two superharmonic functions which are equal almost everywhere are identical.

Superharmonicity for smooth functions is explained as follows:

A C^2-function in an open set G is superharmonic if and only if $\Delta f \leq 0$ in G.

This fact is a very simple consequence of Theorem 5.1.2, and we leave it as an exercise.

A function f which is superharmonic in an open set G is *locally integrable*: As is clear from Figure 6.1, every point $b \in G$ has a neighborhood in which f is integrable. Note that f being lower semicontinuous, it is bounded below on compact subsets of G, and therefore f is integrable in $B(a,r)$ if $f(a) < \infty$.

A superharmonic function can be quite bad as we shall see later. However, it is a very useful fact that it can be approximated by smooth superharmonic functions:

Let f be superharmonic in an open set G and D a relatively compact open subset of G. Let $0 \leq \varphi$ be radial (i.e., $\varphi(x) = \varphi(|x|)$) and C^∞ with support in $B(0,1)$ such that $\int \varphi = 1$. Then, $\varphi_n(x) = n^d \varphi(nx)$ has support in $B(0, 1/n)$ and $\int \varphi_n = 1$. For all large n, $B(0, 1/n) \subset G$ for all $a \in \overline{D}$ and by Fubini,

$$f_n = f * \varphi_n \quad (\text{define } f = 0 \text{ off } G)$$

is superharmonic in D for all large n. Using polar coordinates and superharmonicity of f,

$$f_n \leq f \quad \text{in } \overline{D} \text{ for large } n.$$

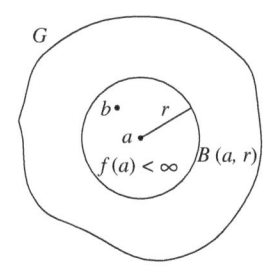

Fig. 6.1 Ball $B(a,r)$ where the superharmonic function f is integrable.

And (as in the proof of (6.1.2)) by lower semi-continuity of f,

$$\lim f_n = f \quad \text{in } \overline{D}.$$

Thus, in each relatively compact open subset D of G, f is a limit of C^∞-functions which are superharmonic in D.

Let us retain the above notation. For large n, f_n are superharmonic in D and C^∞ in $\mathbb{R}^d$. Hence, $\Delta f_n \leq 0$ in D. Let T be the exit time from D and $S_1 \leq S_2$ any stopping times. Proposition 5.1.1 is applicable to $T \wedge S_i$, $i = 1, 2$. Since $\Delta f_n \leq 0$,

$$\mathbb{E}_a \left[f_n \left(X_{T \wedge S_1} \right) \right] - f_n(a) \geq \mathbb{E}_a \left[f_n \left(X_{T \wedge S_2} \right) \right] - f_n(a).$$

Fatou's lemma is applicable since f and hence f_n are uniformly bounded below on $\overline{D}$. Letting n tend to infinity,

$$\mathbb{E}_a \left[f_n \left(X_{T \wedge S_2} \right) \right] \leq \mathbb{E}_a \left[f_n \left(X_{T \wedge S_1} \right) \right], \quad a \in \overline{D}.$$

If also $f \geq 0$, from the last inequality (note that since $S_1 \leq S_2$, the event $(S_2 \geq T) \supset (S_1 \geq T)$),

$$\mathbb{E}_a \left[f \left(X_{S_2} \right) : S_2 < T \right] \leq \mathbb{E}_a \left[f \left(X_{S_1} \right) : S_1 < T \right], \quad a \in \overline{D},$$

which as D increases to G yields

$$\mathbb{E}_a \left[f \left(X_{S_2} \right) : S_2 < R \right] \leq \mathbb{E}_a \left[f \left(X_{S_1} \right) : S_1 < R \right], \quad a \in G,$$

where $R = $ exit time from G.

Let us collect the above in the following:

Theorem 6.1.1. *Let f be superharmonic in G. Then for every relatively compact open subset of G,*

$$\mathbb{E}_a \left[f \left(X_T \right) \right] \leq f(a), \quad T = \text{exit time from } D. \tag{6.1.3}$$

If further $f \geq 0$, then for any stopping times $S_1 \leq S_2$,

$$\mathbb{E}_a \left[f \left(X_{S_2} \right) : S_2 < R \right] \leq \mathbb{E}_a \left[f \left(X_{S_1} \right) : S_1 < R \right] \leq f(a), \quad a \in G, \tag{6.1.4}$$

where $R = $ exit time from G.

The following proposition gives us an alternative and useful definition of superharmonic functions.

Proposition 6.1.2. *Let f be lower semicontinuous in an open set G. Then f is superharmonic in G if and only if it has the following property:*

For every relatively compact open $D \subset G$ and every function u which is continuous on $\overline{D}$ and harmonic in D, $u \leq f$ on ∂D implies $u \leq f$ on D.

Proof. That a superharmonic function satisfies the property follows from the first part of Theorem 6.1.1 and (5.1.5). Conversely, suppose the lower semicontinuous function f satisfies the above condition and let $B(a,r) \subset G$. If g is continuous on $S(a,r)$, its Poisson integral (see Example 5.2.6) is continuous on $B(a,r)$, harmonic on its interior, and $\leq f$ on $S(a,r)$. By our hypothesis,

$$\mathbb{E}_a[g(X_T)] \leq f(a), \quad T = \text{exit time from } B(a,r).$$

Letting g increase to f, we see that f is superharmonic. $\square$

The above proposition easily implies that *superharmonicity is a local property*:

Proposition 6.1.3. *Let $f : G \to (-\infty, \infty]$ be lower semicontinuous. Suppose for each $a \in G$ there exists $r(a)$ such that*

$$\mathbb{E}_a[f(X_{T_r})] \leq f(a)$$

for $r \leq r(a)$, where $T_r = $ exit time from $B(a,r)$. Then f is superharmonic in G.

Proof. Indeed, let u be harmonic in a relatively compact open subset D of G. Suppose u is continuous on $\overline{D}$ and $u \leq f$ on ∂D. $f - u$ is lower semicontinuous on $\overline{D}$. Let α be the minimum of $f - u$ on $\overline{D}$. Now (6.1.1) is valid for f and each a in G provided $r \leq r(a)$. It follows that if $f(a_0) - u(a_0) = \alpha$, for a point a_0 in D, the same holds almost everywhere, and hence everywhere in a neighborhood of a_0, because the set $(f - u = \alpha)$ is closed. Thus, if $f - u = \alpha$ at any interior point a_0 in D, then the same holds in the component of D containing a_0. Since $f - u \geq 0$ on ∂D, we must have $\alpha \geq 0$. This proves the claim. $\square$

Excessive functions. This notion was introduced by G. Hunt. A function $f : G \to [0, \infty]$ is called excessive if

$$\mathbb{E}.[f(X_t) : t < R] \leq f(\cdot), \quad R = \text{exit time from } G$$

and tends to f as t tends to zero.

Markov property easily implies that $\mathbb{E}.\left[f\left(X_t\right):t<R\right]$ increases as t decreases. Now for $t>0$, $\mathbb{E}.\left[f\left(X_t\right):t<R\right]$ is lower semicontinuous being the increasing limit of continuous functions $\mathbb{E}.\left[f\left(X_t\right)\wedge n:t<R\right]$ (Exercise 5.2.3). Since $\mathbb{E}.\left[f\left(X_t\right):t<R\right]$ increases to f as t tends to zero,

every excessive function is lower semicontinuous.

By ordinary Markov property, $f\left(X_t\right)1_{t<R}$ is a super martingale relative to $\mathbb{P}_a$ provided $f(a)<\infty$. Therefore, for discrete stopping times T,

$$\mathbb{E}_a\left[f\left(X_T\right):t<R\right]\leq f\left(a\right). \tag{6.1.5}$$

Since any stopping time is a limit of discrete ones, lower semi-continuity of f ensures that (6.1.5) is valid for any stopping time. Taking $T=$ exit time from $B\left(a,r\right)\subset G$, we get the defining property of super harmonic functions. Note that the case $f(a)=\infty$ is trivial.

If $0\leq f$ is superharmonic, the second part of Theorem 6.1.1 (with $S=t$) and lower semi-continuity of f show that f is excessive. Thus,

$0\leq f$ is hyperharmonic if and only if f is excessive.

Proposition 6.1.4. *Let f be superharmonic in an open set G. Let D be a relatively compact open subset of G. Then*

$$u\left(\cdot\right)=\mathbb{E}.\left[f\left(X_T\right)\right],\quad T=\text{exit time from }D$$

is superharmonic in G and harmonic in D.

Proof. Since $u=f$ off $\overline{D}$, Proposition 6.1.3, it is sufficient to show that f is superharmonic in a neighborhood of $\overline{D}$. f is bounded below in a neighborhood of $\overline{D}$. Replacing G by a neighborhood of $\overline{D}$ if necessary, we may assume that $f\geq 0$ in G and hence excessive. The continuous bounded excessive functions

$$f_n\left(\cdot\right)=\mathbb{E}.\left[\left(f\wedge n\right)\left(X_{\frac{1}{n}}\right):\frac{1}{n}<R\right],\quad R=\text{exit time from }G$$

increase to f. So by monotone convergence, we may further suppose that f is bounded, continuous, and excessive in G. Then $f\left(X_t\right)1_{t<R}$ is a continuous bounded super martingale relative to $\mathbb{P}..$ Since $t+T\left(\theta_t\right)$ decreases to T as t rends to zero, by the Markov property and optional sampling theorem for super martingales,

$$\mathbb{E}_a\left[u\left(X_T\right):t<R\right]=\mathbb{E}_a\left[f\left(X_{t+T(\theta)}\right):t+T\left(\theta\right)<R\right]$$
$$\leq\mathbb{E}_a\left[f\left(X_T\right)\right]=u\left(a\right).$$

That the left side of the above converges to $u(a)$ as t tends to zero follows from bounded convergence and the continuity of f. Thus, u is excessive and hence superharmonic. That it is harmonic in D is clear. □

Remark 6.1.5. In Proposition 6.1.4, suppose f were harmonic in D and superharmonic in G. Can we conclude that $u = f$? We shall see that the answer to this is intimately connected with the problem of uniqueness of solution of the modified Dirichlet problem (Section 5.2).

Example 6.1.6. If u is harmonic in an open set, $|u|^p$ is subharmonic for all $p \geq 1$ as is easily seen using Holder's inequality. If u is subharmonic, $\exp(u)$ is subharmonic. To see this, if u is also C^∞, then $\Delta(e^u) = (\Delta u + |\text{grad} u|^2)e^u \geq 0$. In the general case, u is the decreasing limit (on any compact subset) of a sequence of C^∞ subharmonic functions. More generally, use of Jensen's inequality (Rudin [16, p. 61]) shows that $A(u)$ is subharmonic if u is subharmonic and A is convex and increasing on the line. If u is harmonic, we need only assume that A is convex.

Example 6.1.7. If f is holomorphic in an open set of the plane, $\log|f|$ is harmonic in the open set ($f \neq 0$), as is seen by Cauchy–Riemann equations. By Proposition 6.1.3, $\log|f|$ is subharmonic. It follows that if f and g are analytic, $p, q > 0$, $p \log|f| + q \log|g| = \log|f|^q|g|^q$ is subharmonic. From Example 6.1.6, $|f|^q|g|^q$ is subharmonic. This is a very useful fact in complex function theory.

Taking $f(x) = x$, we see that $-\log|x|$ is superharmonic on $\mathbb{R}^2$ and harmonic off the origin. If $d \geq 3$, the function $|x|^{-d+2}$ is superharmonic off the origin (just differentiate). From Proposition 6.1.3, $|x|^{-d+2}$ is superharmonic on $\mathbb{R}^d$, $d \geq 3$. These are the so-called fundamental harmonic functions. A function which is harmonic and radial in a neighborhood of the origin is of the form $aK(x) + b$, $a > 0$ and b are constants and where

$$K(x) = -\log|x| \quad \text{if } d = 2, \quad K(x) = |x|^{-d+2} \quad \text{if } d \geq 3.$$

Example 6.1.8. There are more subharmonic functions than is evident at first sight. Every polynomial on $\mathbb{R}^d$ is the difference of two subharmonic polynomials. This need only be shown for homogeneous polynomials. According to Exercise 5.5.3, every homogeneous polynomial of degree n is a linear combination of polynomials of the form $\left(\sum_1^d a_i x_i\right)^n$, where a_i are reals. If n is even, $\left(\sum_1^d a_i x_i\right)^n$ is subharmonic since $\sum a_i x_i$ is harmonic. Write $\left(\sum_1^d a_i x_i\right)^{2n+1} = \sum a_i x_i \left(\sum a_j x_j\right)^{2n}$ and note that each of the summands is a difference of two subharmonic polynomials as seen by the formula $2uv = (u+v)^2 - u^2 - v^2$.

Example 6.1.9. We have seen that excessive functions are the same as non-negative hyperharmonic functions. If $d \geq 3$, $K(x) = |x|^{-d+2}$ is excessive on $\mathbb{R}^d$ and hence by restriction, every subdomain of $\mathbb{R}^d$, $d \geq 3$, has non-constant excessive functions. However, there are no non-constant excessive functions on $\mathbb{R}^2$:

Suppose u is excessive on $\mathbb{R}^2$. Let $a, b \in \mathbb{R}^2$, and $S = $ hitting time to $B(b, r)$. Since the two-dimensional Brownian motion is recurrent (Exercise 5.1.3), $\mathbb{P}_a(S < \infty) = 1$. Using (6.1.4),

$$u(a) \geq \mathbb{E}_a[u(X_S)].$$

As r tends to zero, by lower semi-continuity of u, $u(a) \geq u(b)$, i.e., u is constant.

An open subset $G \subset \mathbb{R}^2$ is called Greenian if there is a non-constant excessive function defined on it. Question: Which open subsets of $\mathbb{R}^2$ are Greenian? Kakutani has the answer.

Theorem 6.1.10. *Let G be a domain in $\mathbb{R}^2$ and $T = $ exit time from G. Then $\mathbb{P}_a[T < \infty] \equiv 1$ or $\equiv 0$ in $\mathbb{R}^2$.*

G is Greenian if and only if $\mathbb{P}_a[T < \infty] \equiv 1$.

Proof. Put $u(a) = \mathbb{P}_a[T < \infty]$. u is easily seen to be excessive on $\mathbb{R}^2$, so it is a constant. Clearly, $\mathbb{P}_a[T < \infty, T > n]$ tends to zero as $n \to \infty$ and

$$\mathbb{P}_a[T < \infty, T > n] = \mathbb{E}_a[P_{X_n}(T < \infty) : T > n]$$
$$= c\mathbb{P}_a[T > n] = c(1 - \mathbb{P}_a[T \leq n]),$$

where $c \equiv u$ is a constant. As $n \to \infty$, we get $c(1 - c) = 0$, i.e., $c = 1$ or 0. If $\mathbb{P}_a[T < \infty] \equiv 1$, the function $u(a) = \mathbb{P}_a[T > t]$ is excessive and non-constant on G for each $t > 0$. Indeed, if D_n increase to G with $\overline{D}_n \subset G$ and $T_n = $ exit time from D_n, then $T_n \uparrow T$ and since $T < \infty$, $T(\theta_{T_n}) \downarrow 0$. It follows that $\mathbb{P}_a[T(\theta_{T_n}) > t]$ tends to zero. If $\mathbb{P}_a[T > t]$ were a constant, this constant must be zero which is absurd (why?). On the other hand, let u be a positive superharmonic function on G. Assume $\mathbb{P}_a[T < \infty] \equiv 0$, i.e., the Brownian path starting at any point $a \in \mathbb{R}^2$ immediately enters G and then never exits from G. For any $t > 0$, $X_t \in G$, and we may easily verify that $\mathbb{E}_a[u(X_t)]$ is superharmonic and non-negative on $\mathbb{R}^2$, so it must be a constant. As $t \to 0$, for each $a \in G$, $\mathbb{E}_a[u(X_t)] \to u(a)$. So u is constant. $\square$

Exercises

6.1.1. Let f be hyperharmonic in an open set G. The set of points $a \in G$ such that there exists a neighborhood $U \ni a$ with $\int_U |f| < \infty$ is open and closed in G.

Remark. If G is connected, this implies that $f(a) < \infty$ for one $a \in G$. Then f is locally integrable in G.

6.1.2. An increasing limit of hyperharmonic functions is hyperharmonic.

6.1.3 (The Minimum Principle). Let $0 \leq f$ be superharmonic in a domain G. Then either $f > 0$ or $f \equiv 0$.

Hint: The set $(f = 0)$ is closed. It is also open because by (6.1.1), $f(a) = 0$ implies f is almost everywhere zero in a neighborhood of a.

6.1.4 (Boundary Minimum Principle). Let f be superharmonic in an open set G. Suppose f is bounded below in G and

$$\liminf_{G \ni y \to x} f(y) \geq 0, \quad x \in \partial G.$$

(If G is unbounded, the point at infinity is considered an element of ∂G.) Then $f \geq 0$ in G.

Hint: In the first part of Theorem 6.1.1, let D increase to G and use Fatou.

6.1.5. Show by an example that the $\liminf$ condition in Exercise 6.1.4 cannot be replaced by a $\limsup$ condition.

Hint: Let (a_n) be dense in $S(0,1)$, T = exit time from $B(0,1)$. Then $\mathbb{P}_0[X_T \in (a_n)] = 0$. Let U be open $\supset (a_n)$ such that $\mathbb{P}_0[X_T \in U] < \frac{1}{3}$ say. Let $f = 1$ on U and -1 elsewhere on $S(0,1)$. $u(\cdot) = \mathbb{E}.[f(X_T)]$ is harmonic in $B(0,1)$ and tends to $f(b)$ for all $b \in U$ because f is continuous on U (see Remark 5.2.4). $u(0) < 0$ and $\limsup u = 1$ on $S(0,1)$.

6.1.6. Let G be open and R = exit time from G. For every non-negative Borel function f and every open or closed subset A, the functions

$$\mathbb{E}.\left[\int_0^R f(X_t)\, dt\right] \quad \text{and} \quad \mathbb{P}.[T < R]$$

are excessive in G. T = hitting time to A.

6.1.7. Let m be any measure on $\mathbb{R}^d$, $d \geq 3$. Then $\int |\cdot - y|^{-d+2} m(dy)$ is hyperharmonic on $\mathbb{R}^d$.

Hint: Fatou and Fubini.

6.1.8. Let m be any Radon measure on $\mathbb{R}^2$. Then $v(\cdot) = \int |\log|\cdot - y|| \, m(dy)$ is locally integrable unless $\equiv \infty$. In the former case, $\int \log^+ |x - y| \, m(dy) < \infty$ for all x and $u(\cdot) = \int \log |\cdot - y| \, m(dy)$ is subharmonic on $\mathbb{R}^2$.

Hint: If $v < \infty$ at one point, m is necessarily a finite measure. Using $\log^+ |x + y| < \log 2 + \log^+ |x| + \log^+ |y|$ show that $\int \log^+ |x - y| \, m(dy)$ is (finite and) subharmonic. Since $\log |x - y| \leq \log^+ |x - y|$ and the latter is m-integrable, Fubini and Fatou can be used to show that u is subharmonic.

6.1.9. Let s be superharmonic in a neighborhood of $\overline{D}$, D bounded open. If $a \in \partial D$ is irregular,
$$s(a) = \liminf_{D \ni b \to a} s(b).$$

Hint: s is bounded below in a neighborhood of $\overline{D}$ so assume s is excessive. The Brownian path starting at a must remain in D for a positive length of time. Now use excessivity and lower semi-continuity of s.

6.1.10. Let D be bounded open, $T = $ exit time from D and $a \in \partial D$ irregular. Then there exists a sequence $a_n \in D$, $a_n \to a$, such that the harmonic measures $\mathbb{P}_{a_n}(X_T \in dz)$ tend weakly to the harmonic measure $\mathbb{P}_a(X_T \in dz)$.

Hint: Let B be a ball containing $\overline{D}$. The minimum of two continuous and excessive functions in B is continuous and excessive in B. The differences of functions which are continuous and excessive in B are thus lattices and are therefore uniformly dense in $C(\overline{D})$. It is thus enough to show that there exists a sequence a_n in D converging to a such that $\mathbb{E}_{a_n}[s(X_T)]$ tends to $\mathbb{E}_a[s(X_T)]$ for every s which is continuous and excessive in B. For this, let $\{s_n\}$ be a dense subset of the set
$$J = \{s : 0 \leq s \leq 1, \, s \text{ continuous and excessive in } B\}$$
considered as a subset of the separable Banach space $C(\overline{D})$. Put $s = \sum 2^{-n} s_n$. $u = \mathbb{E}.[s(X_T)]$ is then excessive in B. Using Exercise 6.1.9, let $a_m \in D$ be such that
$$\lim_m u(a_m) = u(a).$$
This implies using semi-continuity
$$\lim_m \mathbb{E}_{a_m}[s_n(X_T)] = \mathbb{E}_a[s_n(X_T)]$$
for every n.

6.2 Applications

The fact that $\log|f|$ is subharmonic for f analytic has many applications in complex function theory. We consider one application in H^p-spaces. For more on this subject, consult Duren [5].

Let f be analytic in the unit disk $B(0,1)$. f is said to be in H^p, $p > 0$, if

$$\sup_{0 < r < 1} \int |f(rz)|^p \, \sigma(dz) < \infty, \tag{6.2.1}$$

where σ is the uniform distribution on $S(0,1)$. The real and imaginary parts of f are harmonic. By the Poisson formula,

$$f(x) = \int_{|y|=1} P_r(x,y) f(ry) \, \sigma(dz), \quad |x| < r, \tag{6.2.2}$$

where

$$P_r(x,y) = \left(r^2 - |x|^2\right) |x-y|^{-2}, \quad |x| < r, \ |y| = 1.$$

Standard proofs of the following theorem use Blaschke products.

Theorem 6.2.1. *Let $p \geq 1$ and $f \in H^p$. There exists $f^* \in L^p(\sigma)$ such that*

$$f(x) = \int P_r(x,y) f^*(y) \, \sigma(dz), \quad |x| < 1. \tag{6.2.3}$$

Proof. We prove this for $p = 1$. For $p > 1$, the argument is similar and simpler. Consider the functions $g_r(z) = |f(rz)|^{\frac{1}{2}}$. Equation (6.2.1) with $p = 1$ implies that the functions g_r form a bounded subset of $L^2(\sigma)$. So there is a sequence $r_n \to 1$ and $g_1 \in L^2$ such that g_{r_n} tends weakly to g_1 in L^2. Recalling $g = |f|^{\frac{1}{2}}$ is subharmonic in $|x| < 1$, from the first part of Theorem 6.1.1 (and Poisson integral formula),

$$g(x) \leq P_{r_n}(x,y) g_{r_n}(y) \sigma(dy), \quad |x| < r_n.$$

By weak convergence, the above inequality holds with r_n replaced by 1. Since $P_1(x,y) \sigma(dy)$ is a probability measure,

$$g^2(x) \leq \int P_1(x,y) g_1^2(y) \sigma(dy). \tag{6.2.4}$$

Integrating both sides of the above on $(|x| = 1)$ (remember $g(rx) = g_r(x)$),

$$\|g_r\|_2 \leq \|g_1\|_2,$$

$\|\cdot\|_2$ denoting L^2-norm. Since $g_{r_n} \to g_1$ weakly, (6.2.4) shows that g_{r_n} converges to g_1 in L^2. But then $g_{r_n}^2$ converges to g_1^2 in L^1, i.e., that $|f(r_n \cdot)|$

is a convergent sequence in L^1 and in particular it is uniformly integrable. Thus, the sequence $f(r_n \cdot)$ is also uniformly integrable. There exists an $f^* \in L^1$ such that a subsequence of $f(r_n \cdot)$ converges weakly to f^*. See Meyer [12, p. 20]. It is now obvious that (6.2.3) holds. □

Remark. The Fatou boundary limit theorem (Section 5.4) and Theorem 6.2.1 show that $f(r \cdot)$ actually converges in L^1 to f^*. If $f \in H^p$ for $0 < p < 1$, it is still true that $f(r \cdot)$ converges in L^p to a function $f^* \in L^p$. This needs a little more work. See the exercises.

The following corollary, which is equivalent to Theorem 6.2.1, is important in prediction theory.

Corollary 6.2.2 (F. and M. Riesz). *Let m be a complex measure on $S(0,1)$ for which $\int z^n m(dz) = 0$ for $n = 1, 2, \ldots$. Then m is absolutely continuous relative to σ.*

Proof. Using the condition, it is easily seen that the Poisson integral of m belongs to H^1. Now use Theorem 6.2.1 and the uniqueness of Poisson integrals. □

Theorem 6.2.3 (F. and R. Nevanlinna). *Let N be the class of functions analytic in the unit disk for which*

$$\sup_{0<r<1} 1(+, r, f) < \infty,$$

where $1(+, r, f)$ is the L^1-norm of $\log^+ |f(r\cdot)|$. $f \in N$ if and only if it is the quotient of two bounded analytic functions.

Proof. Suppose $f \in N$. $w = \log^+ |f|$ is subharmonic in the unit disk. Let T_r denote the exit time from $B(0, r)$. Then

$$w(\cdot) \leq \mathbb{E}.\,[w(X_{T_r})], \quad |x| < r.$$

The right is positive and harmonic for $|x| < r$. It increases as r increases (simply use strong Markov property). The limit u is either $\equiv \infty$ or harmonic. The condition implies that $u(0) < \infty$. Thus, u is a positive harmonic function that dominates w. Let v be conjugate harmonic so that $g = u + iv$ is analytic in the disk. $h = \exp g$ is analytic and $|h| = \exp u > 1$. $|f| \leq h$ because $\log^+ |f| < u$. (f/h) and h^{-1} are both analytic and bounded by 1, $f = (f/h) h$.

Conversely, suppose $f = (a \mid b)$, where a and b are analytic and bounded in the disk. We may assume that $|a| \leq 1$, $|b| \leq 1$ and that $b(0) \neq 0$. Then $\log^+ |f| \leq -\log|b|$:

$$-\infty < \log|b(0)| \leq \mathbb{E}_0\left[\log|b(X_{T_r})|\right]$$
$$= \mathbb{E}_0\left[\log^+ |b(X_{T_r})|\right] - \mathbb{E}_0\left[\log^- |b(X_{T_r})|\right],$$

$\log^+ |b(X_{T_r})| \leq |b(X_{T_r})|$ and b is bounded. Thus, we must have

$$\sup_r \mathbb{E}_0\left[\log|b(X_{T_r})|\right] < \infty. \qquad \square$$

Let G be an open set. A subset A of G is called polar if there is a superharmonic function s on G such that $A \subset s^{-1}(\infty)$. Since s is locally integrable, a polar set is of Lebesgue measure zero. Actually, a polar set is much thinner as we see in the following.

Let F be a relatively closed polar subset of G. Suppose s is superharmonic on $G\backslash F$ and is locally bounded below (this means that for each compact $K \subset G$, s restricted to $K\backslash F$ is bounded below). Extend s in a lower semicontinuous fashion, i.e., define

$$s(a) = \liminf_{G\backslash F \ni y \to a} s(y). \qquad (6.2.5)$$

Clearly, $s : G \to (-\infty, \infty]$ is unaltered on $G\backslash F$. We claim that s is superharmonic on G. To this end, we use Proposition 6.1.2.

Suppose u is harmonic in a relatively compact open set $D \subset G$, continuous on $\overline{D}$, and $u \leq s$ on ∂D.

Let f be superharmonic on G such that $F \subset A = f^{-1}(\infty)$. Since f is bounded below on D, we may assume that $f \geq 0$ on D. For any $\epsilon > 0$, $F = s + \epsilon f$ is identically ∞ on A. Using Proposition 6.1.3, it is clear that F is superharmonic on G. Also $u \leq F$ on ∂D and hence $u \leq F$ on D. Letting ϵ tend to zero, $u \leq s$ on $D\backslash A$. Since A has measure zero, $u \leq s$ on $D\backslash F$ (use (6.1.2)). From (6.2.5), $u \leq s$ on D. Thus, we have the following:

Proposition 6.2.4. *Let F be a relatively closed polar subset of an open set G. If s is superharmonic on $G\backslash F$ and locally bounded below, there is a superharmonic function on G which agrees with s on $G\backslash F$.*

Theorem 6.2.5 (Rado). *Let f be a continuous complex-valued function on an open set G of the plane. Suppose that f is holomorphic at every point of G at which f is not zero. Then f is holomorphic in G.*

Proof. There is no loss of generality in assuming that G is connected. Proposition 6.1.3 implies that $\log|f|$ is subharmonic in G. Therefore, the set $(f=0) = (\log|f| = -\infty)$ is polar. If $f = u + iv$, both u and v must be harmonic in G, i.e., f is holomorphic in G. □

Exercises

6.2.1. Let $0 \neq f \in H^1$ and f^* be as in Theorem 6.2.1. Show that

$$\log|f|(x) \leq \int P_1(x,y) \log|f^*|(y) \sigma(dy), \quad |x| < 1.$$

In particular, $\log|f^*| \in L^1$ and so f^* cannot vanish on a set of positive measure.

Hint: In the proof of Theorem 6.2.1, it was shown that $|f(r_n \cdot)|$ converges to $|f^*|$ in L^1. Assume $|f(r_n \cdot)|$ converges to $|f^*|$, σ-almost everywhere. Using $|z| - \log|z| \geq 0$, and Fatou,

$$\limsup_{r_n \to 1} \int P_{r_n}(x,y) \log|f(ry)| \sigma(dy) \leq \int P_1(x,y) \log|f^*(y)| \sigma(dy)$$

and the terms on the left dominate $\log|f(x)|$ as soon as $|x| < r_n$ because $\log|f|$ is subharmonic. $\log|f^*| \leq |f^*|$ so only $\log^-|f^*|$ can have a divergent integral. Choosing x so that $f(x) \neq 0$, one concludes that $\log|f^*| \in L^1(\sigma)$.

6.2.2. Let $f \in H^p$, $p > 0$. Show that there exists $f^* \in L^1(\sigma)$ such that $f(r \cdot)$ tends to f^* in L^p as r tends to 1.

Hint: f is necessarily in N. $f = g/h$, where h and g are bounded analytic. The radial limits of g and h exist and are non-zero almost everywhere. So the radial limit f^* of f exists almost everywhere and is in L^p. Just as in the proof of Theorem 6.2.1, $|f(r \cdot)|$ converges in L^p to $|f^*|$. Now use Exercise 6.2.3.

6.2.3. Let $f_n \in L^p$, $p > 0$. Suppose f_n converges to f almost everywhere and $\int |f_n|^p$ tends to $\int |f|^p$. Then f_n converges to f in L^p.

Hint: $g_n = |f_n|^p \wedge |f|^p$ is bounded by an integrable function. $|f_n|^p - g_n$ are non-negative and their integrals converge to zero. Thus, $|f_n|^p$ is uniformly integrable. So is $|f_n - f|^p \leq 2^p(|f_n|^p + |f|^p)$.

6.2.4. Show that a countable union of polar sets is polar.

Hint: Suppose $E_n \subset s_n^{-1}(\infty)$, with s_n superharmonic in G. If D_n are open relatively compact with $\bigcup_n D_n = G$, s_n is bounded below on $\overline{D}_n$: say $s_n \geq b_n$ on $\overline{D}_n$. Let $a_n > 0$ such that $\sum a_n |b_n| < \infty$. $s = \sum a_n s_n$ is superharmonic on G: For $n \geq m$, $s_n - b_n$ is non-negative and superharmonic on D_m. $s \equiv \infty$ on $\bigcup E_n$.

6.3 Riesz Measure

We associate, with a superharmonic function f, a measure called its *Riesz measure*. In the language of Schwartz distributions, this measure is simply $(-\Delta) f$. We see that this has important consequences.

Let us start with the simple identity

$$\int F(x) \Delta G(x)\, dx = \int G(x) \Delta F(x)\, dx, \qquad (6.3.1)$$

where G is a C^2-function with compact support and F is C^2 in a neighborhood of the support of G.

Lemma 6.3.1. *Let f be locally integrable in an open set Ω. Suppose for every $F \in C_0^\infty(\Omega)$ (C^∞-functions with compact support contained in Ω)*

$$\int f \Delta F = 0.$$

Then there exists a harmonic function g in Ω such that $f = g$ almost everywhere.

Proof. We show that f is equal almost everywhere (in Ω) to a C^2-function; such a function must be harmonic by (6.3.1).

Let D be a relatively compact open set with $\overline{D} \subset \Omega$. Let V be a neighborhood of 0 such that $\overline{D} + V \subset \Omega$. Let A be any C^∞-function with support contained in V. If F is C^∞ and has support contained in D, for every $x \in V$, $F(x - y)$ as a function of y is C^∞ with support in Ω. From our condition on f,

$$\int f(x-y) \Delta F(y)\, dy = \int f(y) \Delta F(x-y)\, dy = 0$$

for all $x \in V$, implying

$$\int A(x)\, dx \int f(x-y) \Delta F(y)\, dy = \int \Delta F(y)\, dy \int A(x) f(x-y)\, dx = 0.$$

$\int A(x) f(x-y)\, dx$ is C^∞ in D and the above equation holds for all C^∞-functions F with support in D. This implies that $g(y) = \int A(x) f(x-y)\, dx$ is harmonic in D and so has the mean value property.

Let D_1 be open with $\overline{D}_1 \subset D$ and V_1 a neighborhood of 0 for which $\overline{D}_1 + V_1 \subset D$. If B is C^∞ with support in V_1 and B depends only on distance, i.e., $B(x) = B(|x|)$, the mean value property of g implies that $g * B = g$ in D_1, i.e., $A * f * B = A * f$. This holds for all C^∞ functions with support in V implying that $f * B = f$ almost everywhere in D_1. That is to say that f is almost everywhere equal in D_1 to a C^∞-function. That completes the proof. □

Lemma 6.3.2. *Let f be superharmonic in an open set Ω. There exists a unique measure m such that*

$$\int f \Delta \varphi = -\int \varphi \, dm \qquad (6.3.2)$$

for every φ in C^2 with compact support contained in Ω. $m = 0$ if and only if f is harmonic.

Proof. Uniqueness is clear. If $m = 0$, by Lemma 6.3.1, f is equal almost everywhere to a harmonic function, but then f must itself be harmonic. (That two superharmonic functions equal almost everywhere are identical follows from the remarks made just after the definition of a superharmonic function).

f being locally integrable, we can define a linear functional L on the set of C^2-functions with compact support in Ω by

$$LF = -\int f \Delta F.$$

We claim $LF \geq 0$ if $F \geq 0$. To see this, let $D \subset \overline{D} \subset \Omega$ be an open set with compact closure such that the support of F is contained in D. If f_n is the sequence of smooth superharmonic functions we constructed in the beginning of Section 6.1 (to approximate f),

$$LF = -\int f \Delta F = \lim_n -\int f_n \Delta F = \lim_n \int (-\Delta) f_n F \geq 0$$

since $f_n \leq 0$ and $F \geq 0$. By the Riesz representation theorem, a non-negative linear functional on the set of C^2-functions with compact support in Ω is given by a positive Radon measure. □

Definition. The measure m given by Lemma 6.3.2 is called the *Riesz measure of superharmonic function f.*

Note that no claim is made that m is a finite measure; m, however, is finite on compact subsets of Ω. We write down some properties of the Riesz measure m corresponding to a superharmonic function.

Properties of m.

1. For any open subset U of Ω, the Riesz measure of f/U is m/U. This is obvious from (6.3.2). The Riesz measure of $f_1 + f_2$ is $m_1 + m_2$.
2. For any open subset U of Ω, $m(U) = 0$ if and only if f is harmonic in U. Use Property 1 above and (6.3.2).
3. If f_n are superharmonic with Riesz measures m_n, and $f_n \uparrow f$, then m_n tend weakly to m. Indeed, for any C^2-function φ with compact support in Ω,

$$\lim_n \int \varphi dm_n = -\lim_n \int f_n \Delta\varphi = -\int f\Delta\varphi = \int \varphi dm.$$

We now apply Lemma 6.3.2 to a special superharmonic function. Let

$$K(x) = \begin{cases} -|x| & d = 1 \\ -\log|x| & d = 2 \\ |x|^{-d+2} & d \geq 3. \end{cases} \quad (6.3.3)$$

K is superharmonic on $\mathbb{R}^d$ and is harmonic except at the origin. By Property 2, the Riesz measure of K must be concentrated at the origin: For any C^2-function F with compact support,

$$\int K(x)\Delta F(x)\,dx = -A_d F(0), \quad (6.3.4)$$

where A_d is a constant. Taking

$$F(x) = \begin{cases} \left(1 - |x|^2\right)^3 & |x| \leq 1 \\ 0 & |x| > 1 \end{cases}$$

in (6.3.4) gives us the value of A_d:

$$A_d = \begin{cases} 2 & d = 1 \\ 2\pi & d = 2 \\ \frac{(d-2)2\pi^{d/2}}{\Gamma(d/2)} & d \geq 3 \end{cases} \quad (6.3.5)$$

because the area of the unit sphere $\mathbb{R}^d$ is

$$2\pi^{d/2}\Gamma(d/2).$$

If F is C^2 and has compact support, the function $F(x - \cdot)$ has the same properties. From (6.3.4), we get the first part of the following:

Theorem 6.3.3. *For any C^2-function F with compact support,*

$$\int K(y) \Delta F(x-y) \, dy = -A_d F(x). \tag{6.3.6}$$

If f is C^1 with compact support, the function

$$u(x) = \int K(x-y) f(y) \, dy \tag{6.3.7}$$

is C^2 and satisfies the Poisson equation

$$\Delta u = -A_d f. \tag{6.3.8}$$

The constants A_d are given in (6.3.5).

Proof. That u given by (6.3.7) is C^2 if f is C^1 with compact support is routine. To prove (6.3.8), multiply both sides of (6.3.7) by a C^2-function F with compact support, and use (6.3.1), Fubini, and (6.3.6) to get

$$\int F \Delta u = -A_d \int f F.$$

And this is equivalent to (6.3.8). □

Suppose s_1 and s_2 are superharmonic in an open set G with Riesz measures m_1 and m_2. Suppose $m_1 = m_2$ in an open set $D \subset G$. $s_1 - s_2$ is defined almost everywhere in D and is locally integrable. From the definition of Riesz measures and Lemma 6.3.1, there is a function h, harmonic in D such that $s_1 - s_2 = h$, i.e., $s_1 = s_2 + h$ almost everywhere and hence everywhere, because both sides are superharmonic. As a corollary of this observation, we have the following:

Proposition 6.3.4 (Bôcher). *Every function f which is positive and harmonic in the punctured ball $V = B(0,1) \setminus \{0\}$ is of the form*

$$f = cK + h,$$

where the constant $c \geq 0$ and h is harmonic in $B(0,1)$.

Proof. Since $\{0\}$ is clearly polar, by Proposition 6.2.4, there is no loss of generality in assuming that f is superharmonic in $B(0,1)$. Its Riesz measure can only be concentrated at the origin. This follows from the above observation. □

We use this proposition in Chapter 7. See also Exercise 6.3.5.

Theorem 6.3.5 (F. Riesz). *Let f be superharmonic in an open set G with Riesz measure m. To any relatively compact open subset D of G corresponds a function $g(D,\cdot)$ which is harmonic in D and superharmonic in G and satisfies*

$$\frac{1}{A_d}\int_D K(x-y)\, m(dy) + g(D,x) = f(x), \quad x \in G. \qquad (6.3.9)$$

The constants A_d are defined in (6.3.5).

Proof. Using the observation made before Proposition 6.3.4, there is a function $g(D,\cdot)$ harmonic in D such that (6.3.9) holds in D. Off D the first term in (6.3.9) is harmonic. Define $g(D,\cdot)$ off $\overline{D}$ by (6.3.9). To define $g(D,\cdot)$ on ∂D, let D_1 be any open relatively compact set with $D_1 \supset \overline{D}$ and define

$$g(D,x) = g(D_1,x) + \frac{1}{A_d}\int_{D_1\setminus D} K(x-y)\, m(dy), \quad x \in D_1.$$

$g(D,\cdot)$ is then superharmonic in D_1. It is easy to see that this definition of $g(D,\cdot)$ on ∂D does not depend on D_1. □

Remark 6.3.6. If $\int_D K(x_0-y)\, m(dy) < \infty$ for one x_0, then

$$f(x) = \frac{1}{A_d}\int_D K(x-y)\, m(dy) + h \qquad (6.3.10)$$

with h harmonic in G. Indeed, the first term on the right side of the above equality is then superharmonic on $\mathbb{R}^d$ with Riesz measure m. In particular, we see that if m is finite with bounded support, then (6.3.10) holds. This is not true in general. For example, in $\mathbb{R}^2$, the function $f(x) = |x|$ is subharmonic with Riesz measure $-\frac{1}{|y|}dy$ (see the following examples) and $\int |\log|x-y||\frac{1}{|y|}dy = \infty$ for every x.

Example 6.3.7. If u is harmonic, $|u|^p$ is subharmonic for $p \geq 1$. If $p = 1$, the Riesz measure of $|u|$ is clearly concentrated on the set $u = 0$. Using Green's second identity (see, for instance, Buck [3, p. 443]), we can show that the Riesz measure of $|u|$ is a multiple of $\frac{\partial u}{\partial n}d\sigma$, where $d\sigma$ is the surface element on $(u=0)$ and $\frac{\partial u}{\partial n}$ is the normal derivative; the point is that for a harmonic function, level sets (i.e., sets $u = a$, a constant) are not too bad, but we do not go into this. If $p > 1$, the Riesz measure of $|u|^p$ is absolutely continuous relative to the Lebesgue measure with density a multiple of $|u|^{p-2} |\mathrm{grad} u|^2$.

If $f(z)$ is analytic, $|f|^p$ is subharmonic for all $p > 0$. It can be seen using Theorem 6.3.5 that the Riesz measure of $|f|^p$ has no atoms. Outside of its zero set, $|f(z)|^p$ is differentiable so that its Riesz measure has density a multiple of $\Delta |f(z)|^p$. On the other hand, $\log |f|$ is also subharmonic, and clearly its Riesz measure must be concentrated on the zero set of f. If a is a zero of f of multiplicity n, $(z-a)^{-n} f(z)$ is holomorphic at a and does not vanish at a. So the Riesz measure of $\log |f|$ has an atom of weight $2\pi n$ at the point a.

Example 6.3.8. The function K defined in (6.3.3) is superharmonic on $\mathbb{R}^d$. If $B = B(0,1) = \{x : |x| < 1\}$, the function v defined by

$$v(x) = K(x) \wedge K(1) = \begin{cases} K(1) & x \in B \\ K(x) & x \notin B \end{cases}$$

is superharmonic on $\mathbb{R}^d$. v is clearly invariant under rotations. Therefore, the Riesz measure of v (which must of course be concentrated on ∂B) is a multiple of the uniform distribution σ on B. By Remark 6.3.6,

$$v(x) = \frac{c}{A_d} \int_{|y|=1} K(x-y) \sigma(dy) + h, \quad c = \text{constant},$$

where h is harmonic in $\mathbb{R}^d$. But it is also rotation invariant. It must therefore be a constant. (If u is harmonic, $u(x) = f(|x|)$, where f is defined on $(0, \infty)$, $0 = \Delta u(x) = f''(|x|) + \frac{d-1}{|x|} f'(|x|)$, implying for $d \geq 2$ that f is unbounded near 0 unless constant.) For $|x| > 1$, $K(x-z)$ as a function of z is harmonic for $|z| < |x|$. The mean value property gives

$$K(x) = \int K(x-y) \sigma(dy), \quad |x| > 1.$$

All the above facts imply that for $|x| > 1$ (note $v(x) = K(x)$ for $|x| \geq 1$), $\left(1 - \frac{c}{A_d}\right) K(x)$ is equal to a constant, i.e., that the constant is zero and $c = A_d$. Therefore,

$$v(x) = \int_{|y|=1} K(x-y) \sigma(dy), \quad x \in \mathbb{R}^d.$$

σ = uniform distribution on $|y| = 1$. As a simple application, this last equality for $d = 2$ and $x = 1$ clearly implies $\frac{1}{2\pi} \int \log |1 - \cos \theta| \, d\theta = 1$.

The following generalizes Example 6.3.8.

Example 6.3.9. Let f be superharmonic on an open set G and $D \subset \overline{D} \subset G$ bounded open set. The function g defined by

$$g(x) = \begin{cases} f(x) & x \notin \overline{D} \\ \mathbb{E}_x[fX_T] & x \in \overline{D}, \end{cases}$$

where T = exit time from D, is superharmonic in G by Proposition 6.1.4. We claim that the Riesz measure n of g is

$$n(dz) = \int \mathbb{P}_a(X_T \in dz)\, m(da),$$

where m = Riesz measure of f and $\mathbb{P}_a(X_T \in dz)$ = the Dirac measure at a for $a \notin \overline{D}$. We need the following symmetry result which is shown in Chapter 7:

$$\int K(x-y)\,\mathbb{P}_a(X_T \in dz) = \int K(a-y)\,\mathbb{P}_X(X_T \in dz)$$

for all $a, x \in D$. Assuming this, let A be a bounded open set with $\overline{D} \subset A \subset \overline{A} \subset G$, then

$$f(x) = \frac{1}{A_d} \int_A K(x-y)\, m(dy) + h,$$

where h is harmonic in A because m is the Riesz measure of f. Since $\mathbb{E}_a(h(X_T)) = h(a)$ for all $a \in A$,

$$g(x) = \frac{1}{A_d} \int_A m(dy) \int K(z-y)\,\mathbb{P}_X(X_T \in dz) + h$$

$$= \frac{1}{A_d} \int_A K(z-y)\, n(dz) + h,$$

using the above symmetry result. But this simply means that n is the Riesz measure of g as claimed.

Remark. Example 6.3.9 says that the Riesz measure of g is obtained as follows: The measure m is left undisturbed outside $\overline{D}$. The part of m in $\overline{D}$ is "swept tout" onto ∂D by use of the harmonic measures $\mathbb{P}.(X_T \in dy)$. This operation is the "*sweeping out*" or balayage method and n is called the "balayage" of m relative to D. We return to this in Chapter 8.

Exercises

6.3.1. Show that the gradient of a superharmonic function (defined on an open set) exists almost everywhere.

Solution. By Theorem 6.3.5, it is clearly sufficient to show the following:

If m is a positive measure with compact support in $\mathbb{R}^d$, the function

$$u(x) = \int K(x-y)\, m(dy)$$

has gradient almost everywhere.

The case $d = 1$ is simple. If $d \geq 2$, $a < b$ reals, and $x = (x_1, \ldots, x_d)$, $y = (y_1, \ldots, y_d)$,

$$\int_a^b \frac{x_1 - y_1}{|x-y|^d}\, dx_1 = K(b - y_1, x_2 - y_2, \ldots, x_d - y_d)$$
$$- K(a - y_1, x_2 - y_2, \ldots, x_d - y_d)$$

in the sense that if one side is meaningful, the other side is too and equality holds. For any $M < \infty$,

$$\int_{|x| \leq M} dx \int_{|y| \leq N} \frac{|x_1 - y_1|}{|x-y|^d} m(dy) = \int_{|y| \leq N} m(dy) \int_{|x| \leq M} \frac{|x_1 - y_1|}{|x-y|^d}\, dx$$
$$\leq \int_{|y| \leq N} m(dy) \int_{|z| \leq M+N} \frac{1}{|z|^d}\, dz,$$

where the ball $|y| \leq N$ contains the support of m. This is for almost all $(x_2, \ldots, x_d)$ the integrals

$$\int_a^b dx_1 \int \frac{|x_1 - y_1|}{|x-y|^d} m(dy) < \infty.$$

For any such $(x_2, \ldots, x_d)$,

$$\int_a^b dx_1 \int \frac{x_1 - y_1}{|x-y|^d} m(dy) = \int m(dy) \int_a^b \frac{x_1 - y_1}{|x-y|^d}\, dx_1$$
$$= u(b, x_2, \ldots, x_d) - u(a, x_2, \ldots, x_d).$$

This proves the result.

6.3.2. Let u be superharmonic in an open set. If the Riesz measure of u has bounded density, then $u \in C^1$.

Hint: By Theorem 6.3.5, it is enough to show that if g is bounded and measurable with compact support, the function $u(x) = \int K(x-y) g(y) dy$ is continuously differentiable. If φ is locally L^1 and $\psi \in L^\infty$ with compact support, $\varphi * \psi$ is continuous. So the result follows since K and $\operatorname{grad} K$ are locally in L^1.

6.3.3. Let g be Hölder continuous with compact support. Then

$$u(x) = \int K(x-y) g(y) dy$$

is C^2 on $\mathbb{R}^d$. (A function g is Hölder continuous if there are numbers $\alpha > 0$, $M > 0$ such that $|g(x) - g(y)| < M |x-y|^\alpha$.)

Solution. From Exercise 6.3.2, u is continuously differentiable and the partial $u_1 = \frac{\partial u}{\partial x_1}$ is given by

$$u_1(x) = \int \frac{x_1 - y_1}{|x-y|^d} g(y) dy.$$

Let A be a C^1-function with compact support, which is equal to 1 on the support of g. We assume that $|g(x) - g(y)| \leq |x-y|^\alpha$ and $0 \leq A \leq 1$. There is no loss of generality in this. We show that the second partial u_{11} is given by

$$u_{11}(x) = \int \frac{1}{|x-y|^d} (g(x) - g(y)) A(y) dy$$

$$- d \int \frac{(x_1 - y_1)^2}{|x-y|^{d+2}} (g(x) - g(y)) A(y) dy \qquad (*)$$

$$+ g(x) \int \frac{x_1 - y_1}{|x-y|^d} A_1(y) dy,$$

where $A_1 = \frac{\partial A_1}{\partial x_1}$. By the Hölder continuity of g, the function

$$\frac{1}{|x-y|^d} |g(x) - g(y)| A(y) \leq \frac{1}{|x-y|^{d-\alpha}} A(y)$$

and

$$\sup_z \int \frac{1}{|z-y|^{p(d-\alpha)}} A(y)^p dy < \infty \qquad (**)$$

provided $p(d-\alpha) - (d-1) < 1$. Therefore, the integrands

$$\frac{1}{|x-y|^d} (g(x) - g(y)) A(y)$$

are uniformly integrable showing that the first integral on the right side of (∗) is continuous in x. A similar conclusion is obtained with the other two integrals in (∗).

Let e_1 denote the unit vector $(1, 0, 0, \ldots, 0)$. The difference quotient

$$\frac{1}{h}[u_1(x + he_1) - u_1(x)]$$

$$= \frac{1}{h}\int\left(\frac{x_1 + h - y_1}{|x + he_1 - y|^d} - \frac{x_1 - y_1}{|x - y|^d}\right)(g(y) - g(x))A(y)\,dy$$

$$+ g(x)\frac{1}{h}\int\left(\frac{x_1 + h - y_1}{|x + he_1 - y|^d} - \frac{x_1 - y_1}{|x - y|^d}\right)A(y)\,dy. \qquad (***)$$

After a change of variable, the second term on the right side of $(***)$ becomes

$$g(x)\frac{1}{h}\int\frac{y_1}{|y|^d}(A(x + he_1 - y) - A(x - y))\,dy$$

which tends as $h \to 0$ to

$$g(x)\int\frac{y_1}{|y|^d}A_1(x - y)\,dy.$$

To show that lim and $\int$ can be interchanged in the first integral in $(***)$, we show that the integrands in the first term of $(***)$ have a bounded L^p norm if $p(d - \alpha) - (d - 1) < 1$. Write $P = |x + he_1 - y|$, $Q = |x - y|$. Then

$$\frac{1}{|h|}\left|\frac{x_1 + h - y_1}{P^d} - \frac{x_1 - y_1}{Q^d}\right| \le \frac{1}{|h|}\frac{|P^d - Q^d|}{P^{d-1}Q^d} + \frac{1}{Q^d}$$

since $|x_1 + h - y_1| \le P$. $Q^{-d}(g(y) - g(x))A(y)$ is the first integrand on the right side of (∗) and has already been shown to have a bounded L^p-norm. The function

$$\frac{1}{|h|}\frac{|P^d - Q^d|}{P^{d-1}Q^d}|g(y) - g(x)|A(y) \le \frac{1}{|h|}\frac{|P^d - Q^d|}{P^{d-1}Q^{d-\alpha}}A(y) = B(h, x, y)$$

say; $|P^d - Q^d| \le |P - Q| \cdot d \cdot (\max(P, Q))^{d-1}$ and $|P - Q| \le |h|$. On the set $P \ge Q$, $B(h, x, y) \le dQ^{-d+\alpha}A(y)$. And on the set $P < Q$, $B(h, xy) \le P^{1-d}A(y)Q^{\alpha-1} \le P^{-d+\alpha}A(y)$. And in both cases, (∗∗) can be applied. That completes the proof.

For a simpler proof using Greens identity, see Courant–Hilbert [17].

6.3.4. Let G be open and E a polar subset of G. Show that E is a polar subset of $\mathbb{R}^d$.

Hint: Use Theorem 6.3.5 and Exercise 6.2.4.

6.3.5. Let D be open and F a compact polar subset of D. If u is positive harmonic in $D\backslash F$, then for a measure m on F, $u(x) = \int K(x-y) m(dy) + h(x)$, where h is harmonic in D.

Hint: Same as Proposition 6.3.4.

6.4 The Continuity Principle

Superharmonic functions can be very discontinuous. For example, let $\{a_n\}$ be a sequence dense in $\mathbb{R}^3$. We can choose positive numbers b_n such that

$$u(x) = \sum b_n \frac{1}{|x-a_n|}$$

defines a superharmonic function on $\mathbb{R}^3$. (One need only choose b_n so that the series converges at one point.) $u(a_n) = \infty$ a for every n so that u is discontinuous at every point at which it is finite. Thus (recall u is lower semicontinuous), the set of points of continuity of u (continuity in the generalized sense) coincides with the set of infinities of u. This last set has measure zero. On the other hand, u being lower semicontinuous, the set of points of continuity of u is a dense G_δ-set.

Thus, it would appear that little can be said about the continuity properties of superharmonic functions in general. Theorem 6.3.5 shows that local properties of superharmonic functions can be investigated with the help of superharmonic functions of the form $\int K(x-y) m(dy)$, where m is a finite measure with compact support. For example, the gradient of a superharmonic function exists almost everywhere; see Exercise 6.3.1. See also Exercise 6.3.2.

The following theorem is very important.

Theorem 6.4.1 (Continuity Principle of Evans-Vasciliesco). *Let f be superharmonic in an open set G. If the restriction of f to the support of its Riesz measure is continuous, then f is continuous on G.*

Proof. By Theorem 6.3.5, it is sufficient to show the following: Let m be a finite measure with compact support F. If the restriction of

$$u(x) = \int K(x-y) m(dy) \qquad (6.4.1)$$

to F is finite and continuous, then u is continuous on $\mathbb{R}^d$.

Let $x_0 \in F$. For each neighborhood V of x_0,
$$u(x) = \int_V + \int_{F\setminus V} = s_V + s_{F\setminus V}, \qquad (6.4.2)$$
say. Both the summands on the right side of (6.4.2) are continuous on F because each is lower semicontinuous and their sum is continuous on F. Also as V decreases to x_0, S_V decreases to zero because m has no atom at x_0 ($u(x) < \infty$). Given $\epsilon > 0$, using Dini, we can find a V such that
$$\sup_{x \in F} s_V(x) < \varepsilon, \quad m(V) < \epsilon. \qquad (6.4.3)$$
For any x, let z_x be a point in F nearest to x:
$$|z_x - y| \le |z_x - x| + |x - y| \le 2|x - y|, \quad y \in F$$
so that for all $y \in F$,
$$K(x - y) \le \log 2 + K(z_x - y), \qquad d = 2,$$
$$K(x - y) \le 2^{d-2} K(z_x - y), \qquad d \ge 3.$$
Therefore, for all x,
$$s_V(x) \le \log 2 \cdot m(V) + s_V(z_x), \qquad d = 2,$$
$$s_V(x) \le 2^{d-2} s_V(z_x), \qquad d \ge 3.$$
In other words, using (6.4.3),
$$\sup s_V(x) \quad \text{is small if } V \text{ is small}. \qquad (6.4.4)$$
Thus, since $s_{F\setminus V}$ is continuous at x_0,
$$\limsup_{x \to x_0} u(x) \le \sup_x s_V(x) + s_{F\setminus V}(x_0)$$
$$\le \sup_x s_V(x) + u(x_0)$$
which, using (6.4.4), gives the continuity of u at x_0. □

Theorem 6.4.2. *Let G be a domain and f excessive in G, with Riesz measure m. Suppose f is finite m-almost everywhere. Then*
$$f = \sum_n f_n + h, \qquad (6.4.5)$$
where f_n are finite continuous and excessive and h is harmonic and ≥ 0.

Proof. Let D_n be an increasing sequence of open relatively compact sets with union G. Let us show that there exists a continuous excessive function f_1 (on G) whose Riesz measure m_1 is concentrated on D_1 such that

$$f - f_1 \text{ is excessive in } G \text{ and } m(D_1) - m_1(D_1) < \frac{1}{2}.$$

By Theorem 6.3.5,

$$s_1(x) + g(D_1, x) = f(x), \quad x \in G,$$

where

$$s_1(x) = \frac{1}{A_d} \int_{D_1} K(x-y) \, m(dy).$$

Since f is m-almost everywhere finite, $s_1 < \infty$ m-almost everywhere. By Lusin's theorem (Rudin [16, p. 53]), we can find a closed set $F \subset D_1$ such that s_1/F is continuous and $m(D_1 \backslash F) < \frac{1}{2}$. Now

$$s_1 = \frac{1}{A_d} \int_F + \frac{1}{A_d} \int_{D_1 \backslash F}.$$

Since s_1/F is continuous and each of the summands above is lower continuous, $s_2(x) = \frac{1}{A_d} \int_F K(x-y) \, m(dy)$ is continuous on F and hence on $\mathbb{R}^d$ by the continuity principle:

$$s_2(x) + u(x) = f(x), \quad x \in G,$$

$$u(x) = \frac{1}{A_d} \int_{D_1 \backslash F} K(x-y) \, m(dy) + g(D_1, x), \quad x \in G. \quad (6.4.6)$$

Let T_n = exit time from D_n. Since u is superharmonic,

$$s_2(x) - \mathbb{E}_x[s_2(X_{T_n})] \le f(x) - \mathbb{E}_x[s_2(X_{T_n})] \le f(x). \quad (6.4.7)$$

$\mathbb{E}_x[s_2(X_{T_n})]$ is harmonic in D_n and decreases as n increases. The limit says h_2 is either $\equiv -\infty$ or is harmonic in G. The former is excluded by (6.4.7). Writing $f_1 = s_2 - h_2$, f_1 is excessive and continuous, its Riesz measure m_1 is concentrated on D_1, $m(D_1) - m_1(D_1) < 2^{-1}$, and $0 \le f - f_1 = u + h_2$ is excessive in G. Repeat the above argument with $f - f_1$ and D_2 to get a continuous excessive function f_2 whose Riesz measure m_2 is concentrated on D_2 such that

$$f - f_1 - f_2 \text{ is excessive in } G,$$
$$m(D_2) - m_1(D_2) - m_2(D_2) < 2^{-2}.$$

In general, we can find a continuous function f_n with Riesz measure m_n concentrated on D_n such that

$$f - \sum_{1}^{n} f_i \text{ is excessive in } G,$$

$$m(D_n) - \sum_{1}^{n} m_i(D_n) < 2^{-n}.$$

Clearly, $\sum f_i$ is excessive in G with Riesz measure m. Equation (6.4.5) then follows from the observations made after Theorem 6.3.3. □

Remark. The simple example $f(x) = K(x)$ shows that the assumptions in the above theorem are not superfluous.

Exercises

6.4.1. Suppose u is given by (6.4.1) where m has support F. If the restriction of u to F is continuous at $x_0 \in F$, then u is continuous at x_0.

Hint: Using continuity of u/F at x_0, find a neighborhood V of x_0 such that

$$s_V(x) = \int_V K(x-y) m(dy)$$

and $m(V)$ are small for all $x \in V$. Then use (6.4.4).

6.4.2. Let L be a kernel on $\mathbb{R}^d$, i.e., L is defined on $\mathbb{R}^d \times \mathbb{R}^d$, $L > 0$, $L(x,x) \equiv \infty$, and $(x,y) \mapsto L(x,y)^{-1}$ is finite and continuous. Suppose $F \subset \mathbb{R}^d$ is compact and m a measure on F such that $Lm(x) = \int L(x,y) m(dy)$ is continuous when restricted to F. If Lm is not continuous, then there exists a measure p on F such that $Lp \leq 1$ on F and $\sup_x Lp(x) = \infty$.

Hint: Suppose $x_0 \in F$ and $\limsup_{x \to x_0} Lm(x) > \epsilon + Lm(x_0)$. Using Dini find neighborhoods V_n of x_0 such that $m(V_n) < 2^{-2n}$, $Lm_n \leq 2^{-2n}$ on F, where $m_n = m/V_n$. We must have $\limsup_{x \to x_0} Lm_n(x) > \epsilon + Lm_n(x_0)$. $p = \sum 2^n m_n$ is then the required measure.

6.5 The Dirichlet Problem Revisited

Let D be a bounded open set. For any continuous function f on ∂D, the function $u(a) = \mathbb{E}_a[f(X_T)]$, $T =$ exit time from D, is harmonic in D and $\lim_{b \to a} u(b) = f(a)$ for every regular point $a \in \partial D$. This we have seen in Chapter 5. It is far from obvious that u is uniquely determined by these properties. We show in this section that this is indeed the case.

Lemma 6.5.1. *Let G be open and D an open relatively compact subset of G. Let f be superharmonic in G with Riesz measure m. If f is finite m-almost everywhere and harmonic in D then*

$$f(a) = \mathbb{E}_a[f(X_T)], \quad T = \text{exit time from } D. \qquad (6.5.1)$$

Proof. Since f is bounded below in a neighborhood of $\overline{D}$, we may (by replacing G by this neighborhood if needed) assume that f is excessive in G. Now appeal to Theorem 6.4.2 and note that for each n, f_n is continuous on G and harmonic in D so that (6.5.1) is true for each n by (5.1.5). □

Remark. The example $G = \mathbb{R}^3$, $D = \{x : 0 < |x| < 1\}$ and $f(x) = |x|^{-1}$, shows that Lemma 6.5.1 above is not valid in general.

Lemma 6.5.2. *Let D be a bounded open set and $T =$ exit time from D. For every $a \in \overline{D}$, then*

$$\mathbb{P}_a(T(\theta_T) = 0) = 1.$$

In other words, with probability 1, X_T is regular.

Proof. Let B be a ball containing $\overline{D}$ and $S =$ exit time from B. $\mathbb{E}.[S]$ is bounded and excessive in B:

$$v(a) = \mathbb{E}_a[\mathbb{E}_{X_T}(S)] = \mathbb{E}_a[S] - \mathbb{E}_a[T] \qquad (6.5.2)$$

is also bounded and excessive in B and it is harmonic in D. By Lemma 6.5.1, $v(a) = \mathbb{E}_a[V(X_T)]$, which, using (6.5.2), gives $\mathbb{E}_a[T(\theta_T)] = 0$. □

The following theorem contains as a special case the uniqueness of the solution of the modified Dirichlet problem.

Theorem 6.5.3 (The Strong Boundary Minimum Principle). *Let D be a bounded open set. If u is superharmonic in D, is bounded below, and*

$$\liminf_{D \ni b \to a} u(b) \geq 0, \quad a \in \partial D \text{ regular.}$$

Then $u > 0$ in D.

Proof. Let $a \in D$ and $a \in D_n$ be open sets whose closures are contained in D and which increase to D. $T_n =$ exit time from D_n increases to $T =$ exit time from D, $\mathbb{P}_a$-almost surely:

$$u(a) \geq \mathbb{E}_a\left[u(X_{T_n})\right].$$

An application of Fatou's Lemma and Lemma 6.5.2 gives the result. $\square$

Remark. The example $D = \{x \in \mathbb{R}^3 : 0 < |x| < 1\}$ and $u(x) = 1 - \frac{1}{|x|}$, $x \in D$, shows that the boundedness assumption in the above theorem is not superfluous. We have all the machinery now to prove the famous Kellogg–Evans theorem: The set of irregular points in ∂D is polar. See exercises. We return to this in Chapter 8.

Let us consider another application of Theorem 6.4.2. Let $B = B(0, R)$. The minimum of two continuous excessive functions on B is continuous and excessive. It follows that the difference $s_1 - s_2$ with s_1 and s_2 continuous and excessive on B is a lattice and hence by the Stone–Weierstrass theorem, any continuous function on any compact subset of B can be approximated on the set arbitrarily closely by such differences. Let now D be bounded open, $\overline{D} \subset B$, and f continuous on ∂D. For any $\epsilon > 0$, we can find continuous excessive functions s_1 and s_2 such that $s = s_1 - s_2$:

$$|f(x) - s(x)| < \epsilon, \quad x \in \partial D.$$

If $T =$ exit time from D,

$$|\mathbb{E}_a\left[f(X_T) - s(X_T)\right]| < \epsilon, \quad a \in \overline{D}.$$

By Theorem 6.4.2,

$$\mathbb{E}.\left[s_i(X_T)\right] = \sum_n s_{i,n}, \quad i = 1, 2,$$

where $s_{i,n}$ are continuous, excessive (in B), and harmonic in D. If $a \in \partial D$ is regular, $\mathbb{E}_a\left[f(X_{T_n})\right] = f(a)$. Since the partial sums $\sum_{n \leq N} s_{i,n}$ are continuous on $\overline{D}$ and harmonic in D and $s_1 - s_2$ is close to f on ∂D, we have proved the following:

Proposition 6.5.4. *Let D be bounded and open. Each continuous function f on D is the bounded pointwise limit on $r(D)$ ($=$ the set of regular points in ∂D) of a sequence of functions f_n each of which is continuous on $\overline{D}$ and harmonic in D.*

One comment on Proposition 6.5.4 is in order. For unexplained terminology, consult Phelps [14]. One associates with each compact set $X \subset \mathbb{R}^d$ a subspace:

$D(X) = $ set of function in $C(X)$ which are harmonic in the interior of X.

The content of Proposition 6.5.4 together with Lemma 6.5.2 is then as follows:

The Choquet boundary of $D(X) = $ set of points in ∂X that are regular for the complement of the interior of X.

$D(X)$ is simplicial, namely that the only signed measure living on the Choquet boundary and annihilating $D(X)$ is the zero measure.

A positive measure on ∂X is said to represent $x \in X$ if $\int f(y) m(dy) = f(x)$ for every $f \in D(X)$. That $D(X)$ is simplicial is equivalent to saying that each point $x \in X$ has a unique representing measure living on the Choquet boundary. We can say more: For each × in the interior of X, there is only one measure for ∂X which represents x. We return to this point in Chapter 8.

We end this section by giving one last application of Theorem 6.4.2 to a theorem of J.L. Doob.

Theorem 6.5.5 (J.L. Doob). *Let s be superharmonic on an open set G and $T = $ exit time from G. Then for every $a \in G$, $\mathbb{P}_a$-almost surely $s(X_t)$ is continuous in the generalized sense on $0 \le t < T$ and is finite for $0 < t < T$.*

Proof. It is sufficient to prove this theorem by assuming s is excessive; indeed, consider first an open relatively compact subset D of G and then let D increase to G. Assume also s is finite. By Theorem 6.4.2, $s = \sum s_i$, where s_i is continuous excessive in G. $\sum_n^{n+p} s_i(X_t) 1_{t<T}$ is then a right continuous supermartingale, which is continuous for $t < T$. By the supermartingale inequalities of Chapter 2,

$$\epsilon \mathbb{P}_a \left[\sup_{0 \le t < T} \sum_n^{n+p} s_k(X_t) > \epsilon \right] \le \sum_n^{n+p} s_k(a) \le \sum_n^{\infty} s_k(a).$$

This shows the uniform convergence of $\sum_1^n s_k(X_t)$ to $\sum_1^\infty s_k(X_t)$ in the interval $0 \le t < T$.

Now let s be any excessive function. We claim for every $t_0 > 0$ and $a \in G$,

$$\mathbb{E}_a \left[s\left(X_{t_0}\right) : t_0 < T \right] < \infty. \tag{6.5.3}$$

Indeed, let $a \in D$ be any open relatively compact subset of G and $S =$ exit time from D. From (6.1.4),

$$\mathbb{E}_a \left[s\left(X_{t_0 \vee S}\right) : t_0 < T \right] \leq \mathbb{E}_a \left[s\left(X_S\right) \right] < \infty \tag{6.5.4}$$

because from Proposition 6.1.4, $\mathbb{E}. \left[s\left(X_S\right) \right]$ is harmonic in D. Now for $t_0 > 0$,

$$\mathbb{P}_a \left[X_{t_0} \in dy, \, t_0 < S \right] \leq \mathbb{P}_a \left[X_{t_0} \in dy \right] \leq (2\pi t_0)^{-d/2} dy$$

so that

$$\mathbb{E}_a \left[s\left(X_{t_0}\right) : t_0 < T \right] \leq (2\pi t_0)^{-d/2} \int_D s(y) \, dy < \infty \tag{6.5.5}$$

because s is locally integrable. Equation (6.5.3) is a consequence of (6.5.4) and (6.5.5).

Now $s(X_t) 1_{t<T}$ is a non-negative supermartingale. Since s is lower semicontinuous, $(s > n)$ is open. Hence,

$$\mathbb{P}_a \left[s(X_t) > n \text{ for some } t_0 < t < T \right]$$
$$= \mathbb{P}_a \left[s(X_r) > n \text{ for some rational } r \text{ with } t_0 < r < T \right]$$
$$\leq \frac{1}{n} \mathbb{E}_a \left[s(X_{t_0}) : t_0 < T \right]$$

which implies that

$$\mathbb{P}_a \left[(s(X_t) = (s \wedge n)(X_t) \quad \text{for all } t_0 < t < T \right] \quad \text{for some } n] = 1.$$

We let the reader complete the rest. □

Exercises

6.5.1. Let D be a bounded open set. Show that the set of irregular points in ∂D is polar. (*This is the famous Kellogg–Evans theorem.*)

Hint: By Exercise 5.2.4, the set of irregular points in ∂D is a countable union of compact sets. Therefore, by Exercise 6.2.4, it is enough to show that a compact subset F of the set of irregular points is polar. Let B be a ball containing $\overline{D}$ and put $G = B \backslash F$. Then $F \subset \partial G$ and clearly every point in F is irregular for the complement of G. By Lemma 6.5.2, starting in G, it is impossible to hit F. Let V_n be a sequence of open sets with $V_n \supset \overline{V}_{n+1}$ and $\bigcap V_n = F$. Let $R_n =$ hitting time to $\overline{V}_n$. The functions

$$p_n(a) = \mathbb{P}_a(R_n < S), \quad S = \text{exit time from } B$$

are excessive in B, $p_n(a) = 1$ for $a \in V_n$ and $p_n(a)$ decreases to zero for every a in G, because $X_{R_n} \in \overline{V}_n$ on the set $R_n < S$ and therefore a path in $\bigcap_n (R_n < S)$ must hit F. Let $a_0 \in G$. By choosing a subsequence if necessary, we may assume $\sum p_n(a_0) < \infty$. $p = \sum p_n$ is then excessive in B and infinite on F.

References

[1] H. Aikawa. Potential theoretic notions related to integrability of superharmonic functions and supertemperatures. *Anal. Math. Phys.*, 9(2): 711–728, 2019.

[2] M. Alakhrass. Multiply superharmonic functions outside negligible sets. *Ann. Mat. Pura Appl. (4)*, 195(4): 1373–1381, 2016.

[3] R. C. Buck. *Advanced Calculus*. McGraw-Hill Book Co., Inc., New York, 1956.

[4] J. L. Doob. Generalized sweeping-out and probability. *J. Funct. Anal.*, 2: 207–225, 1968.

[5] P. L. Duren. *Theory of H^p Spaces*. Pure and Applied Mathematics, Vol. 38. Academic Press, New York, 1970.

[6] M. El Kadiri, A. Aslimani, and S. Haddad. On the integral representation of the nonnegative superharmonic functions in a balayage space. *Riv. Math. Univ. Parma (N.S.)*, 10(1): 1–24, 2019.

[7] K. Lai, J. Mu, and H. Wang. Retraction note: New applications of Schrödingerean Green potential to boundary behaviors of superharmonic functions. *Bound. Value Probl.*, 1, 2021. Paper No. 65.

[8] J. Liu, S. Shi, and Y. Zhang. Superharmonic functions related to convexity estimates for a class of semilinear elliptic problems. *Nonlinear Anal.*, 236: 12, 2023. Paper No. 113353.

[9] K. Luan and J. Vieira. Poisson-type inequalities for growth properties of positive superharmonic functions. *J. Inequal. Appl.*, 10, 2017. Paper No. 12.

[10] K. Luan and J. Vieira. Retraction note: Poisson-type inequalities for growth properties of positive superharmonic functions. *J. Inequal. Appl.*, 1, 2021. Paper No. 148.

[11] S. Ma and J. Qing. On n-superharmonic functions and some geometric applications. *Calc. Var. Partial Differ. Equ.*, 60(6): 42, 2021. Paper No. 234.

[12] P.-A. Meyer. *Probability and Potentials*. Blaisdell Publishing Co. [Ginn and Co.], Waltham, 1966.

[13] Y. Miura. Superharmonic functions of Schrödinger operators and Hardy inequalities. *J. Math. Soc. Jpn.*, 71(3): 689–708, 2019.

[14] R. R. Phelps. *Lectures on Choquet's Theorem*. D. Van Nostrand Co., Inc., Princeton, 1966.
[15] T. Rado. *Subharmonic Functions*. Ergebnisse der Mathematik und ihrer Grenzgebiete, Vol. 5. Springer-Verlag, Berlin, 1937.
[16] Walter Rudin. *Real and Complex Analysis*. McGraw-Hill Book Co., New York, 1966.
[17] R. Courant and D. Hilbert. *Methods of Mathematical Physics*, Vol. II, Interscience Publishers (a division of John Wiley & Sons, Inc.), New York, 1962, pp. 248–250.

Chapter 7

Green Functions, Boundary Value Problems, and Kernels

The contexts of the results in Chapters 5 and 6 are now connected to a systematic and modern analysis Green functions.

The original material in Chapter 6 dealing with Green functions is still current, and we have added a discussion of new directions. In addition, for the benefit of readers, we offer the following citations covering new directions [4, 6, 7, 10–14].

The new directions include Weyl–Titchmarsh M-functions, Sturm–Liouville operators, and Green's functions; use of Green's function for solving initial-boundary value problems in PDEs; Green and Martin kernels for Schrödinger operators with singular potential in Lipschitz domains; and fractional conformal Laplacians.

For the benefit of beginner readers, we offer the following supplementary introductory text [5].

Introduction. Consider a domain G and the inhomogeneous equation $\Delta u = f$, called the *Poisson equation*. The solution is not unique without further conditions: If u is a solution, so is $u + w$ for any harmonic function w. If G is bounded and we impose the "boundary condition" that u vanish on the boundary, the maximum principle guarantees that there can be at most one solution. In simple terms, the Green function for G is the "inverse of Δ together with this boundary condition". In order to better understand what follows, recall the definition and properties of Green functions for ordinary differential operators (see, e.g., Birkhoff [1, pp. 39–47]).

We see that knowing the Green function for a domain is equivalent (theoretically) to solving the Dirichlet problem.

In Section 7.1, we define Green functions and point out some of their most important properties. Sections 7.2 and 7.3 are devoted to a discussion of Green functions for unbounded open sets in $\mathbb{R}^2$. Section 7.4 contains some examples, and, in Section 7.5, an expression for the Green function in terms of relative transition densities is given.

7.1 Green Functions for Bounded Open Sets

Let D be a bounded open set and $K(x)$ be defined as in (6.3.3). By Theorem 6.3.3, for any C^1-function f with compact support

$$F(x) = \int K(x-y) f(y) \, dy \qquad (7.1.1)$$

is a solution of the Poisson equation

$$\Delta F = -A_d f, \qquad (7.1.2)$$

where A_d are constants defined in (6.3.5). This solution, however, does not satisfy our "boundary condition" of tending to zero at the boundary. In order to secure this, we solve the Dirichlet problem with boundary values F and subtract it from F; namely, if we define

$$u(x) = F(x) - \mathbb{E}_x [F(X_T)], \qquad (7.1.3)$$

where T = exit time from D, then u satisfies $\Delta F = -A_d f$ in D and $u(b)$ tends to zero as b tends to any regular point in ∂D. From what we have seen in Chapters 5 and 6, one cannot expect more. Equation (7.1.3) can be rewritten as

$$u(x) = \int G_D(x, y) f(y) \, dy, \qquad (7.1.4)$$

where

$$G_D(x, y) = K(x, y) - \mathbb{E}_x [K(X_T - y)], \qquad (7.1.5)$$

T = exit time from D.

Caution. The right side of (7.1.5) does not make sense if, e.g., $x = y \in \partial D$ and regular for D^c. However, for each x, $K(x - \cdot)$ and $\mathbb{E}_x [K(X_T - \cdot)]$ are both superharmonic on $\mathbb{R}^d$ and hence locally integrable. Therefore, even if the right side of (7.1.5) does not make sense for some x and y, for each fixed x, the right side of (7.1.4) makes sense for every bounded Borel measurable f with compact support. Note, however, that for each $x \in D$, the right side of (7.1.5) is well defined for all $y \in \mathbb{R}^d$; indeed, for $y \in D$, $\mathbb{E}_x[K(X_T - y)]$ is clearly finite, and for any y, $\mathbb{E}_x[K(X_T - y)] \leq K(x - y)$.

Proposition 7.1.1. *For each $x_0 \in D$, $\mathbb{E}_{x_0}[K(X_T - y)] = K(x_0 - y)$ for almost all $y \notin D$.*

Proof. To show this, let $x_0 \in D_n$, where D_n is open with $\overline{D}_n \subset D_{n+1} \subset D$ and $\bigcup D_n = D$. Let T_n = exit time from D_n. Then $\mathbb{P}_{x_0}[T_n \uparrow T] = 1$, where T = exit time from D. The functions s_n defined by

$$s_n(y) = \mathbb{E}_{x_0}[K(X_{T_n} - y)], \quad y \in \mathbb{R}^d, \tag{7.1.6}$$

are superharmonic on $\mathbb{R}^d$ and decrease because

$$\mathbb{E}_{x_0}[K(X_{T_{n+1}} - y)] = \mathbb{E}_{x_0}[\mathbb{E}_{X_{T_n}}[K(X_{T_{n+1}} - y)]] \leq \mathbb{E}_{x_0}[K(X_{T_n} - y)].$$

If

$$s(y) = \lim s_n(y) \tag{7.1.7}$$

and f is bounded Borel with compact support, then

$$\int s(y) f(y) \, dy = \lim_n \int s_n(y) f(y) \, dy$$

$$= \lim_n \mathbb{E}_{x_0}\left[\int K(X_{T_n} - y) f(y) \, dy\right]$$

(Fubini's theorem is applicable)

$$= \mathbb{E}_{x_0}\left[\int K(X_T - y) f(y) \, dy\right]$$

$$\left(\text{since } \int K(z - y) f(y) \, dy \text{ is continuous on}\right.$$

$$\left. \mathbb{R}^d \text{ by Exercise 6.3.2}\right)$$

$$= \int f(y) \mathbb{E}_{x_0}[K(X_T - y)] \, dy$$

proving that $s(y) = \mathbb{E}_{x_0}[K(X_T - y)]$ for almost all y. Now $y \notin D$ implies that $y \notin \overline{D}_n$, which in turn implies that $s_n(y) = K(x_0 - y)$ for all n, i.e., $s(y) = K(x_0 - y)$, $y \notin D$. Thus, $K(x_0 - y) = \mathbb{E}_{x_0}[K(X_T - y)]$ for almost all y not in D, as claimed. □

This shows that as far as (7.1.4) is concerned, for each $x \in D$ we can safely redefine $G_D(x,y) = 0$ for all $y \notin D$ and we can state the following:

> For any C^1-function with compact support (support not necessarily in D), the function u defined in D by
> $$u(x) = \int_D G_D(x,y) f(y) \, dy, \quad x \in D$$
> satisfies $\Delta u = -A_d f$ in D and
> $$\lim_{x \to b} u(x) = 0, \quad b \in \partial D, \ b \text{ regular for } D^c.$$

$G(x,y)$ defined on $D \times D$ by (7.1.5) is called the *Green function* for the bounded open set D and we define for convenience $G_D = 0$ outside $D \times D$.

The following is a short list of properties of G_D:

Property 1. $G_D(\cdot,\cdot)$ *is excessive in D in each variable when the other is fixed.* $G_D(x_0, \cdot) - K(x_0, \cdot)$ *is harmonic in D for each $x_0 \in D$.*

Property 2. *Let D be connected open. For each $y \in D$ and $b \in \partial D$,*
$$\limsup_{D \ni x \to b} G_D(x,y) = 0$$
if and only if b is regular for D.

Proof. $\mathbb{E}.[K(X_T - y)]$ is by Proposition 6.1.4 superharmonic on $\mathbb{R}^d$ (in particular, lower semicontinuous), $\leq K(\cdot - y)$, and equals $K(b-y)$ for each $b \in \partial D$ that is regular. In the other direction, we can show the following:

Let h be positive superharmonic in an open set D and $b \in \partial D$. If
$$\lim_{D \ni x \to b} h(x) = 0,$$
then b is regular for D^c.

Indeed, assuming b is not regular, we can find $t > 0$ small enough so that $\mathbb{P}_b(T > t) > 0$, where T is the exit time from D. Consider the process $h(X_t) 1_{t<T}$. If $s < t$, on the set $t < T$, $t = s + T(\theta_s)$. By Markov property,
$$\mathbb{E}_b[h(X_t) 1_{t<T}] = \mathbb{E}_b[\mathbb{E}_{X_s}[h(X_{t-s}) 1_{t-s<T}] : s < T].$$

$h > 0$ is superharmonic on D, i.e., h is excessive on D, i.e., for each $a \in D$ and each $r > 0$, $\mathbb{E}_b[h(X_t) : t < T] \leq h(a)$. For $s < T$, $X_s \in D$, we get
$$\mathbb{E}_b[h(X_t) : t < T] \leq \mathbb{E}_b[h(X_s) : s < T].$$

As $s \downarrow 0$, $X_s \to b$ and $h(X_s) \to 0$ (no loss of generality in assuming h is bounded) giving
$$\mathbb{E}_b[h(X_t) : t < T] = 0.$$
This is a contradiction and b therefore must be regular. □

A function h is a *barrier* at $x \in \partial D$ if h is defined, positive, and superharmonic on $V \cap D$ for some neighborhood V of x and $\lim_{V \cap D \ni y \to x} h(y) = 0$. Since regularity is a local property, we have in fact shown above that for any open set D, $b \in \partial D$ is *regular* if and only if there is a barrier at b. (If b is regular for D, V a bounded neighborhood of b, b is regular for $V \cap D$. If S is the exit time from $V \cap D$, $\mathbb{E}_x(S)$ is positive superharmonic on $V \cap D$ and tends to zero as $x \to b$.)

Property 3. *If $U \subset V$ are bounded open, then $G_U \leq G_V$.*

Proof. Indeed, the first exit time from U is less or equal to the first exit time from V and (6.1.4) applies. (In the case of $d \leq 2$, we add a constant to K to make it excessive in a ball containing the closure of V.) □

Property 4. *If D is bounded open and D_n are open and increase yo D, then $G_{D_n}(x, y)$ increases to $G_D(x, y)$ for all $(x, y) \in D \times D$.*

Proof. This is clear from (7.1.5) since the exit times from D_n increase to the exit time from D. □

Property 5. *Let D be bounded open and $y_0 \in D$. If v is non-negative superharmonic on D and is the sum of $K(\cdot - y_0)$ and a superharmonic function, then $v(x) \geq G_D(x, y_0)$ for all $x \in D$.*

Proof. Indeed, let $B \ni x$ be an open set whose closure is contained in D, and let S be the exit time from B. Writing $v = K(\cdot - y_0) + u$, the non-negativity of v implies
$$\mathbb{E}_x[K(X_S - y_0) + u(X_S)] \geq 0,$$
i.e., (u being superharmonic) $u(x) \geq \mathbb{E}_x[u(X_S)] \geq -\mathbb{E}_x[K(X_S - y_0)]$. Therefore, $v(x) \geq K(x - y_0) - \mathbb{E}_x[K(X_S - y_0)]$, which implies, by letting B increase to D, that $v(x) \geq G_D(x, y_0)$. □

Property 6. *Let D be bounded open. G_D is symmetric on $D \times D$:*
$$G_D(x, y) = G_D(y, x), \quad (x, y) \in D \times D. \tag{7.1.8}$$
More generally,
$$\mathbb{E}_x[K(X_T - y)] = \mathbb{E}_y[K(X_T - x)], \quad (x, y) \in \mathbb{R}^d \times \mathbb{R}^d. \tag{7.1.9}$$

Proof. Denote the left side of (7.1.9) by $u(x,y)$ so that the right side is $u(y,x)$. Fix $x_0 \in D$. $u(x_0, \cdot)$ and $u(\cdot, x_0)$ are both superharmonic on $\mathbb{R}^d$ and harmonic in D. If G is an open set with $\overline{G} \subset D$ and S = exit time from G, then

$$u(x_0, y) = \mathbb{E}_y [u(x_0, X_S)] \leq \mathbb{E}_y [K(x_0 - X_S)], \quad y \in G$$

because $u(x_0, z) \leq K(x_0 - z)$ for all z. Letting G increase to D, we obtain by bounded convergence

$$u(x_0, y) \leq \mathbb{E}_y [K(x_0 - X_T)] = u(y, x_0), \quad y \in D.$$

Since $x_0 \in D$ is arbitrary, this last inequality implies (7.1.9) for all $(x,y) \in D \times D$.

We now show that (7.1.9) holds for all $(x,y) \in \mathbb{R}^d \times \mathbb{R}^d$. By Proposition 7.1.1, $u(x_0, y) = K(x_0 - y) \geq u(y, x_0)$ for almost all $y \notin D$. We have just shown that $u(x_0, y) = u(y, x_0)$ for $y \in D$. Thus, $u(x_0, y) \geq u(y, x_0)$ for almost all y and by superharmonicity for all y:

$$u(x_0, y) \geq u(y, x_0), \quad y \in \mathbb{R}^d, \ x_0 \in D.$$

If $x \notin D$ or $x \in \partial D$ and is regular, $u(x,y) = K(x-y) \geq u(y,x)$. Thus, for all y and all x, except perhaps for $x \in \partial D$ and irregular,

$$u(x,y) \geq u(y,x), \quad y \in \mathbb{R}^d, \ x \in \partial D, \ x \text{ not irregular.}$$

We shall presently show that the set of irregular points in ∂D has Lebesgue measure zero. Since for any y, $u(\cdot, y)$ and $u(y, \cdot)$ are both superharmonic, the last inequality holding off the set of irregular points in ∂D implies $u(x,y) \geq u(y,x)$ for all x. Thus, $u(x,y) \geq u(y,x)$ for all (x,y) which is (7.1.9).

It remains to show that the set of irregular points in ∂D has Lebesgue measure zero. It is enough to show this assuming that D is connected. Indeed, every irregular point is an irregular boundary point of a connected component of D (see Exercise 5.2.5) and there are only countably many connected components. If $x_0 \in D$,

$$h(y) = K(x_0 - y) - \mathbb{E}_{x_0} [K(X_T - y)]$$

is superharmonic and strictly positive in D. For $b \in \partial D$,

$$\limsup_{D \ni y \to b} h(y) = 0 \quad \text{iff } b \text{ is irregular}$$

as we saw in Property 2. But from Proposition 7.1.1 for almost all $b \in \partial D$,
$$K(x_0 - b) = \mathbb{E}_{x_0}\left[K(X_T - b)\right] \le \liminf_{D \ni y \to b} \mathbb{E}_{x_0}\left[K(X_T - y)\right]$$
by lower semi-continuity. Thus, for almost all $b \in \partial D$,
$$\limsup_{D \ni y \to b} h(y) = 0.$$
Thus, (7.1.9) is completely established. □

Property 7. *Let D be bounded, V open $\subset D$, $U \subset D\setminus \overline{V}$, and $S = $ exit time from U. Then for all $x \in V$, $y \in U$,*
$$G_D(x, y) = \mathbb{E}_y\left[G_D(X_S, x) : S < T\right]. \qquad (7.1.10)$$
In particular, if $x \in V$ and $D \ni y \notin \overline{V}$,
$$G_D(x, y) \le \sup_{z \in D \cap \partial V} G_D(x, z).$$

Proof. Indeed, for $x \in V$ and $y \in U$, $K(x - y) = \mathbb{E}_y\left[K(X_S - x)\right]$, and so
$$G_D(x, y) = K(x - y) - \mathbb{E}_y\left[K(X_T - x)\right]$$
$$= \mathbb{E}_y\left\{K(X_S - x) - \mathbb{E}_{X_S}\left[K(X_T - x)\right] : S < T\right\}$$
$$= \mathbb{E}_y\left[G_D(X_S, x) : S < T\right].$$
If $U = D\setminus\overline{V}$, for $S < T$, $X_S \in D \cap \partial V$ and the last inequality claimed above follows. □

Now let us briefly see how the knowledge of the Green function for D (a bounded open set) allows us to solve, theoretically, the generalized Dirichlet problem.

Since any function in $C(\partial D)$ can be approximated uniformly on ∂D by a C^∞-function with compact support, it is sufficient to know the solution of the generalized Dirichlet problem with C^∞-boundary values. For a C^∞-function f with compact support,
$$u(x) = f(x) - \frac{1}{A_d} \int_D G_D(x, y) \Delta f(y)\, dy, \qquad x \in D,$$
is the solution of the generalized Dirichlet problem with boundary function f. If furthermore ∂D is smooth, by an application of Green's identity (recall $G_D(x, \cdot) = -A_d \delta_x$, δ_x being the unit mass at x),
$$u(x) = \frac{1}{A_d} \int_{\partial D} D_n G_D(x, y) \Delta f(y)\, \sigma(dy), \qquad x \in D,$$
where σ is the area measure on ∂D and D_n the normal derivative.

Unbounded open sets. We have now seen that the Green function for a bounded open set is related to the Dirichlet problem and the Poisson equation. The uniqueness of solutions in these considerations was a consequence of the maximum principles. In the formulation and deduction of maximum principles for bounded open sets, the compactness of $\overline{D}$ was a key factor. For an unbounded open set D, compactness of $\overline{D}$ requires the addition of the point at infinity. If the point at infinity is considered an element of ∂D, its role has to be investigated. If D is an unbounded open set in $\mathbb{R}^d$ and ∂D is compact, Proposition 7.1.2 shows that an arbitrary positive harmonic function in D has a limit at infinity and in particular is bounded at infinity, provided $d \geq 3$. This is false when $d = 2$ as shown by the example $D = \{x \in \mathbb{R}^2 : |x| > 1\}$ and $u(x) = \log|x|$, $x \in D$. Proposition 7.1.3, on the other hand, shows that for an arbitrary open set $D \subset \mathbb{R}^2$ and any harmonic function in D which is *bounded and continuous* on $\overline{D}$, the formula

$$u(x) = \mathbb{E}_x[u(X_T)], \quad x \in \overline{D}, \ T = \text{exit time from } D$$

holds; in particular, u harmonic in D, bounded and continuous on $\overline{D}$, $u = 0$ on ∂D, imply $u \equiv 0$. Such a conclusion is invalid in $d \geq 3$ dimensions if D is unbounded as shown by the example $D = \{x \in \mathbb{R}^d : |x| > 1\}$ and $u(x) = 1 - K(x)$, $x \in D$. Thus, in considerations of the Dirichlet problem for unbounded open sets, the dimension plays an important role.

Proposition 7.1.2. *Let u be positive and harmonic in $D = \{x \in \mathbb{R}^d : |x| > 1\}$. If $d \geq 3$, $\lim_{x \to \infty} u(x)$ exists. If $d = 2$,*

$$u(x) = c\log|x| + h(x), \quad c \geq 0,$$

where h is positive harmonic in D and $\lim_{x \to \infty} h(x)$ exists. Further, if $d = 2$ and u is positive superharmonic in D and $\lim_{x \to \infty} u(x) = 0$, then $u \equiv 0$.

Proof. The Kelvin transformation relative to $S(0,1) = \partial B(0,1)$ gives a function which is positive and harmonic in the ball $B(0,1)$ punctured at 0. Use a result of Bocher (Proposition 6.3.4) and note that Kelvin transformation is idempotent. Recall also that a positive superharmonic function vanishing at a point vanishes in its connected component. □

Proposition 7.1.3. *Let $D \subset \mathbb{R}^2$ be open. If u is continuous and bounded on $\overline{D}$ and harmonic in D, then $u(x) = \mathbb{E}_x[u(X_T)]$, $x \in \overline{D}$, where $T = $ exit* time from D. In $d \geq 3$ dimensions, the same conclusion holds, provided we assume, in addition, that $u(x)$ tends to zero as $D \ni x \to \infty$.

Proof. Recall that $\mathbb{P}.(T < \infty) \equiv 1$ or $\equiv 0$ on $\mathbb{R}^2$, by Theorem 6.1.10, and in the latter case every bounded (indeed positive) harmonic function on D is a constant. Put

$$D_n = D \cap \{x : |x| < n\}.$$

u is harmonic on the bounded open set D_n and continuous on $\overline{D}_n$. So

$$u(x) = \mathbb{E}_x[u(X_{T_n})], \quad x \in \overline{D}, \quad T_n = \text{exit time from } D_n.$$

If $x \in \overline{D}$, then $x \in \overline{D}_n$ from and after some n. Clearly, $T_n = R_n \wedge T$, where R_n is the exit time from the disc $B_n = \{x : |x| < n\}$. On the set $T_n < T$, $R_n < T$. $\mathbb{P}_x(R_n \uparrow \infty) = 1$, $\mathbb{P}_x(T < \infty) = 1$, u bounded imply, by letting $n \to \infty$, in

$$u(x) = \mathbb{E}_x[u(X_{T_n})] = \mathbb{E}_x[u(X_T) : T = T_n] + \mathbb{E}_x[u(X_{T_n}) : R_n < T]$$

that

$$u(x) = \mathbb{E}_x[u(X_T)], \quad x \in \overline{D}.$$

This proves the proposition. □

Let us return to the discussion of the Green functions for unbounded open sets. One would like the Green function for an unbounded open set to satisfy as many of Properties 1–7 as possible. Property 4 or 5 can be used for this purpose. We use Property 4. Let D be open and D_n an increasing sequence of bounded open sets with union D. By Property 3, $\lim G_{D_n}$ exists for all $(x, y) \in D \times D$. Now if D_n increase to D and A is a bounded open subset of D, then $D_n \cap A$ increases to A. By Properties 3 and 4,

$$\lim G_{D_n} \geq \lim G_{D_n \cap A} \geq G_A$$

showing that the limit $\lim G_{D_n}$ is independent of the exhausting sequence $\{D_n\}$. In $d \geq 3$ dimensions, $K \geq 0$ and a glance at (7.1.5) shows that for $(x, y) \in D \times D$,

$$G_D(x, y) = \lim G_{D_n}(x, y) = K(x, y) - \mathbb{E}_x[K(X_T - y)], \quad (7.1.11)$$

where T = exit time from D (when $T = \infty$, $K(X_T - y) = 0$ by definition). The situation when $d = 2$ is more complicated. In this case, as the following proposition shows, $\lim G_{D_n}(x, y) < \infty$ off the diagonal of $D \times D$ if, and only if, it is finite at one point. *When this limit is finite, we say that D has a Green function* and

$$G_D(x, y) = \lim G_{D_n}(x, y), \quad (x, y) \in D \times D,$$

is its Green function. Whenever we write G_D, we assume that D is Greenian, i.e., that D has a Green function. By Property 6, G_D is symmetric on $D \times D$.

Proposition 7.1.4. *Let D be open $\subset \mathbb{R}^2$ and D_n be bounded open and increase to D. If T_n = exit time from D_n, then*

$$s(x,y) = \lim \mathbb{E}_x[K(X_{T_n} - y)], \quad x, y \in D,$$

is either $\equiv -\infty$ or is separately harmonic in D.

Proof. Suppose first that D is connected. Since $s(\cdot, \cdot)$ is independent of the exhausting sequence $\{D_n\}$, we may assume that D_n are connected and increase to D. For each m, $\mathbb{E}_x[K(X_{T_n} - y)]$ is separately harmonic in the connected set D_m for all $n \geq m$. And they decrease. The limit is either $\equiv -\infty$ or is separately harmonic in D_m and m is arbitrary.

Now suppose D is not connected. If x and y belong to different components of D, they belong to different components of D_n, say A_n, B_n, for large n. $K(\cdot - y)$ is then harmonic in A_n and continuous on $\overline{A}_n$ so that

$$\mathbb{E}_x[K(X_{T_n} - y)] = K(x - y) \quad \text{for all large } n,$$

i.e., $s(x,y) = K(x,y)$. The case x, y in the same component is to be treated.

If D is not connected, the closure of any component of D cannot be $\mathbb{R}^2$. So suppose D itself is connected and $\overline{D} \neq \mathbb{R}^2$. Let $z \notin \overline{D}$ and $x, y \in D_1$, say, where D_n increase to D. Clearly, there is a positive number α such that $|z - \xi| \cdot |y - \xi|^{-1} \geq \alpha$ as ξ varies on ∂D_n, uniformly in n. Hence,

$$\mathbb{E}_x[K(X_{T_n} - y)] - \mathbb{E}_x[K(X_{T_n} - z)] \geq \log \alpha \tag{7.1.12}$$

uniformly in n. Since $K(z - \cdot)$ is harmonic in D_n and continuous on its closure, the second term in (7.1.12) is just $K(z-x)$. Thus, $\mathbb{E}_x[K(X_{T_n} - y)]$ are bounded below by a harmonic function in D. $s(\cdot, \cdot)$ is therefore separately harmonic in D. $\square$

Properties. The function G_D, defined on $D \times D$ by

$$G_D(x,y) = \lim G_{D_n}(x,y), \quad x, y \in D,$$

where D_n is any sequence of bounded open sets increasing to D, clearly satisfies Properties 1, 3, 4, 5, 6, and 7. That it also satisfies Property 2 is shown as follows:

Let $a \in \partial D$ be regular, $x_0 \in D$. We must show that $G_D(x_0, y)$ tends to zero as $y \to a$. Let V be a ball centered at x, with closure contained in D, and $U = D \cap \{x : |x - a| < r\}$. For small r, U will not intersect $\overline{V}$. From (7.1.10),
$$G_D(x_0, y) = \mathbb{E}_y[G_D(X_S, x_0) : S < T], \quad y \in U,$$
where S = exit time from U. $G_D(x_0, \cdot)$ is continuous on $\partial V \subset D$. By the second statement of Property 7, for all $y \notin V$, $G_D(x_0, y) \leq \sup_{z \in \partial V} G_D(x_0, z) = M$, say. Therefore, $G_D(x_0, y) \leq M \mathbb{P}_y[S < T]$, for $y \in U$. If S_1 = exit time from the ball $\{x : |x - a| < r\}$, then $S = S_1 \wedge T$ and $\mathbb{P}_y[S < T] = \mathbb{P}_y[S_1 < T]$. It is thus enough to show that $\mathbb{P}_y[S_1 < T]$ tends to zero as $y \to a$. But this is simple. Indeed, given $\epsilon > 0$, for small t,
$$\mathbb{P}_y[S_1 \leq t] < \epsilon, \quad \text{for all } y \in B(a, r/2).$$
And
$$\mathbb{P}_y[S_1 < T] \leq \mathbb{P}_y[S_1 < t] + \mathbb{P}_y[t < T].$$
As $y \to a$, $\mathbb{P}_y[t < T]$ tends to zero because a is regular.

> Thus in Properties 1 through 7, D can be any open set having a Green function.

As to the Poisson equation, let D be open and unbounded $\subset \mathbb{R}^d$, $d \geq 3$, and f be C^1 on $\mathbb{R}^d$ with compact support. Then there is one and only one C^2-function u defined on D such that $\Delta u = -A_d f$ in D, $u(x)$ tends to zero as x tends to any regular point in ∂D, and $u(x)$ tends to zero as x tends to infinity. And this function u is given by
$$u(x) = \int_D G_D(x, y) f(y) \, dy, \quad x \in D.$$
The uniqueness claim above follows from Theorem 6.5.3.

The discussion of the Poisson equation for unbounded open subsets (having Green functions) of $\mathbb{R}^2$ seems a little more involved. If $D \subset \mathbb{R}^2$ has a Green function and f is C^1 on $\mathbb{R}^2$ with compact support, it is possible to show the following:

There is one and only one bounded C^2-function u defined on D such that $\Delta u = -A_d f$ in D and $u(x)$ tends to zero as x tends to any regular point in ∂D. And this function u is given by
$$u(x) = \int_D G_D(x, y) f(y) \, dy, \quad x \in D.$$
We discuss this in Section 7.3.

The following is a characterization of Green sets. See also Theorem 6.1.10.

Proposition 7.1.5. *An open set $\Omega \subset \mathbb{R}^2$ is Greenian if and only if there exists a non-constant excessive function on Ω.*

Proof. If Ω is Greenian, for any $x_0 \in \Omega$, $G_\Omega(x_0, \cdot)$ is non-constant excessive on Ω. Conversely, suppose h is a non-constant excessive function on Ω. By considering $\sqrt{h}$ if necessary, we may assume that the Riesz measure m of h is not zero. Let D be a relatively compact open subset of Ω. By Theorem 6.3.5,

$$\frac{1}{A_d} \int_D K(x-y)\, m(dy) + g(D, x) = h(x), \quad x \in \Omega. \tag{7.1.13}$$

If $T = $ exit time from D, we get from (7.1.13) (recall $g(D, \cdot)$ is superharmonic in Ω)

$$\frac{1}{A_d} \int_D (K(x-y) - \mathbb{E}_x[K(X_T - y)])\, m(dy) \le h(x) - \mathbb{E}_x[h(X_T)] \le h(x). \tag{7.1.14}$$

As D increases to Ω, the integrand in (7.1.14) increases to G_Ω:

$$\frac{1}{A_d} \int_D G_\Omega(x, y)\, m(dy) \le h(x). \tag{7.1.15}$$

For any x_0 with $h(x_0) < \infty$, since $m \not\equiv 0$, (7.1.15) shows that $G_\Omega(x, \cdot) < \infty$ m-almost everywhere. Thus, Ω is Greenian by Proposition 7.1.4. $\square$

Let Ω be a Greenian open set in $\mathbb{R}^d$ (if $d \ge 3$, this simply means an open subset). $G_\Omega(\cdot, \cdot)$ is separately excessive in Ω. It follows that for any measure m,

$$G_\Omega m(x) = \int G_\Omega m(x, y)\, m(dy) \tag{7.1.16}$$

is excessive in Ω. If Ω is connected, $G_\Omega m$ is either superharmonic or is ∞ in Ω. An easy consequence of Proposition 7.1.4 is

$$\text{The Riesz measure of } \frac{1}{A_d} G_\Omega m \text{ is } m. \tag{7.1.17}$$

Indeed, suppose $\varphi \in C^\infty$ and has compact support in Ω. On the support of φ, $G_\Omega(x, \cdot)$ and $K(x - \cdot)$ are both integrable. Also $G_\Omega(x, y) = K(x - y)$

$- s(x, y)$, where $s(\cdot, \cdot)$ is separately harmonic in Ω, by Proposition 7.1.4:

$$\int (G_\Omega m) \Delta\varphi$$

$$= \int m(dx) \int K(x-y) \Delta\varphi(y) dy - \int m(dx) \int s(x,y) \Delta\varphi(y) dy$$

$$= \int m(dx) \int K(x-y) \Delta\varphi(y) dy = -A_d \int \varphi(x) m(dx) \qquad (7.1.18)$$

because $\int s(x,y) \Delta\varphi(y) dy = 0$. Equation (7.1.18) implies (7.1.17).

Suppose now h is excessive in Ω and has Riesz measure m. As we saw in Proposition 7.1.5, (7.1.15) is valid. From (7.1.17), $h = \frac{1}{A_d} G_\Omega m + u$, where $0 \leq u$ is harmonic in Ω. We have thus

Theorem 7.1.6 (F. Riesz). *Let Ω be a Greenian open subset of $\mathbb{R}^d$. If h is excessive in Ω and has Riesz measure m, then*

$$h = \frac{1}{A_d} G_\Omega m + u,$$

where $0 \leq u$ is harmonic in Ω and $G_\Omega m$ is defined by (7.1.16).

Kelvin transformation and Green functions. Consider inversion relative to the sphere of radius ρ center 0: $S(0,\rho) = \partial B(0,\rho)$. If $x \neq 0$, its inverse $x^* = \rho^2 \frac{x}{|x|^2}$. Let D be open $\subset \mathbb{R}^d \setminus \{0\}$ and D_1 its image under inversion. If f is C^2 on D_1 and g is its Kelvin transform,

$$g(x) = \frac{\rho^{d-2}}{|x|^{d-2}} f(x^*),$$

then

$$(\Delta g)(x) = \frac{\rho^{d+2}}{|x|^{d+2}} (\Delta f)(x^*).$$

This (together with the approximation of superharmonic functions by smooth ones, Section 6.1) shows that Kelvin transformation preserves superharmonicity. Let $x_0 \in D$. The function

$$u(x) = \frac{\rho^{d-2}}{|x|^{d-2}} G_{D_1}(x_0^*, x^*)$$

is positive superharmonic on D. Since $G_{D_1}(x_0^*, y) - K(x_0^* - y)$ is harmonic for $y \in D_1$, its Kelvin transform is harmonic in D:

$$u(x) - \frac{\rho^{d-2}}{|x|^{d-2}} K(x_0^* - x^*)$$

is harmonic in D. The relation $|x_0^* - x^*| = \frac{\rho^2}{|x_0||x|} |x - x_0|$ shows that

$$u(x) = \begin{cases} u(x) - \frac{\rho^{d-2}}{|x|^{d-2}} K(x_0^* - x^*) + \frac{|x_0|^{d-2}}{\rho^{d-2}} K(x_0 - x), & d \geq 3, \\ u(x) - K(x_0^* - x^*) + K(x_0 - x) + \log|x_0||x| - \log \rho^2, & d = 2. \end{cases}$$

Thus, the function $\frac{\rho^{d-2}}{|x_0|^{d-2}} u(x)$, which is positive and superharmonic in D, is the sum of $K(x_0 - x)$ and a superharmonic function. By Proposition 7.1.5, one concludes that

$$\frac{\rho^{d-2}}{|x_0|^{d-2}} u(x) \geq G_D(x_0, x).$$

The Kelvin transformation being idempotent

$$G_D(x_0, x) = \frac{\rho^{2(d-2)}}{|x|^{d-2} |x_0|^{d-2}} G_{D_1}(x_0^*, x^*).$$

Green functions and holomorphic transformations. If $w(x) = u(x) + iv(x)$ is a holomorphic map on an open set D onto an open set D_1, f is C^2 in D_1 and $g(x) = f(w(x))$, then

$$\Delta g(x) = |\operatorname{grad} u(x)|^2 (\Delta f)(w(x))$$

as is seen by using Cauchy–Riemann equations. This, as in the case of the Kelvin transformation, shows that whenever s is superharmonic on D_1, the composed function $h(x) = s(w(x))$ is superharmonic on D. D_1 has a Green function thus implies that D has a Green function by Proposition 7.1.5. Suppose then that D_1 has a Green function. For any $z_0 \in D$,

$$a(z) = G_{D_1}(w(z_0), w(z))$$

is non-negative superharmonic on D. Since

$$G_{D_1}(w(z_0), x) + \log|w(z_0) - x|$$

is harmonic on D_1, so is

$$b(z) = a(z) + \log|w(z_0) - w(z)|$$

on D. We have
$$a(z) = b(z) - \log\left|\frac{w(z_0) - w(z)}{z - z_0}\right| + K(z_0 - z).$$

$\frac{w(z_0)-w(z)}{z-z_0}$ being holomorphic on D, its logarithm is subharmonic in D. The positive superharmonic function a is thus the sum of a superharmonic function and $K(z_0 - z)$. By Property 5, $a(z) \geq G_D(z_0, x)$. In particular, if w is simple, i.e., 1-1 holomorphic from D onto D_1, then
$$G_{D_1}(w(z_0), w(z)) = G_D(z_0, z),$$
i.e., *Green function is a 1-1 conformal invariant.*

Exercises

7.1.1. Let D be a bounded open set and T = exit time from D. Then
$$L(x, y) = \mathbb{E}_x[K(X_T - y)]$$
is lower semicontinuous on $\mathbb{R}^d \times \mathbb{R}^d$.

7.1.2. Let D be bounded open in $\mathbb{R}^d$, u positive harmonic in D, and $z_0 \in \partial D$ regular. Assume that $u(z) \leq O(K(z - z_0)) + O(1)$ and $\lim_{z \to b} u(z) = 0$ for all $z_0 \neq b \in \partial D$ regular. Then $u \equiv 0$.

Hint: Let $x_0 \in D$. By (7.1.9), $\mathbb{E}_{x_0}[K(X_T - z_0)] = K(x_0 - z_0)$. Let $\overline{D}_n \subset D$ increase to D and T_n = exit time from D_n, then
$$K(x_0 - z_0) = \mathbb{E}_{x_0}[K(X_{T_n} - z_0)] \geq \mathbb{E}_{x_0}[K(X_T - z_0)] = K(x_0 - z_0).$$
Since $K(X_{T_n} - z_0)$ tends to $K(X_T - z_0)$ and all the functions are bounded below, $K(X_{T_n} - z_0)$ thus converges in L^1 (relative to $\mathbb{P}_{x_0}$) to $K(X_T - z_0)$. In particular, $K(X_{T_n} - z_0)$, and hence also $u(X_{T_n})$, is uniformly integrable. Now $u(x_0) = \mathbb{E}_{x_0}[u(X_{T_n})]$ and X_{T_n} tends to X_T. Now use Lemma 6.5.2.

Remark. The example D = the unit disc in $\mathbb{R}^2$ punctured at the origin, $u = K$, shows that the restriction in Exercise 7.1.2, that $z_0 \in \partial D$ is regular, is not redundant.

7.1.3. Let D be an unbounded open set in $\mathbb{R}^d$. Suppose u is bounded harmonic in D and $\lim_{D \ni x \to b} u(x) = 0$ for every $b \in \partial D$ that is regular. If $d = 2$, then $u \equiv 0$. If $d \geq 3$ and further u tends to zero at infinity, then $u \equiv 0$.

7.1.4. Let D be Greenian and s superharmonic in D. Suppose the Riesz measure of s is finite. Then

$$s(x) = \int G(x,y)\, m(dy) + h, \quad h \text{ harmonic in } D.$$

Hint: $G(x, \cdot)$ behaves like $K(x - \cdot)$ near x and $G(x, \cdot)$ is bounded off any neighborhood of x (Property 7). Thus, $\int G(x,y)\, m(dy)$ cannot be identically infinite in any connected component of D.

7.2 Unbounded Open Subsets of $\mathbb{R}^2$

We have defined the Green function for unbounded open subsets of $\mathbb{R}^2$ as a limit of Green functions for bounded open sets. Even in very simple cases, it is difficult to apply this definition to find Green functions. By far the most powerful method is to find a "mapping function" to map a given domain D onto another domain D_1 whose Green function is known. This will then allow us to find the Green function for D.

In this and the following section, we investigate the extent to which "formula" (7.1.5) holds; it does not hold in general. For example, if $D = \{x : |x| > 1\}$,

$$G_D(x,y) = \mathbb{E}_x\left[\log|X_T - y|\right] - \log|x - y| + \log|x|,$$

T = exit time from D, as is seen by taking $D_n = \{x : 1 < |x| < n\}$ and using (5.1.9). We find a necessary and sufficient condition for formula (7.1.5) to hold but first some propositions. The discussion also throws some light on the role of the point at infinity.

Until further notice, D denote a Greenian open set, i.e., an open set having a Green function. Given D, let $D_n = B_n \cap D$, where B_n is the disc of radius n, center 0: $B_n = \{x : |x| < n\}$. T_n = exit time from $\overline{D}_n$ and T = exit time from D. Put

$$s_n(x,y) = \mathbb{E}_x\left[K(X_{T_n} - y)\right], \quad x, y \in \mathbb{R}^2. \qquad (7.2.1)$$

$s_n(x, \cdot)$ are superharmonic on $\mathbb{R}^2$ harmonic in D_n and decrease. Let

$$s(x,y) = \lim s_n(x,y). \qquad (7.2.2)$$

Since $s_n(x,y) \leq K(x-y)$, $s(x,y) \leq K(x-y)$. So $s(x,y) < \infty$ unless $x = y$. $S(x,y)$ may *a priori* be $-\infty$. s is symmetric because s_n are (Property 6, Section 7.1).

Suppose $x_0 \in D$, $y_0 \in D$. D being Greenian, $s(x_0, y_0) > -\infty$. From an n on $|X_{T_n} - y_0|$ is clearly bounded away from zero (since $X_{T_n} \in \partial D_n$) and $K(X_{T_n} - y_0)$ is then bounded above. Fatou's lemma is applicable:

$$-\infty < s(x_0, y_0) \le \mathbb{E}_{x_0}[\limsup K(X_{T_n} - y_0)]$$
$$= \mathbb{E}_{x_0}[K(X_T - y_0)].$$

$K(X_T - y_0)$ being bounded above, $\mathbb{E}_{x_0}[K(X_T - y_0)]$ makes sense, and from the last inequality, we conclude that $\mathbb{E}_{x_0}[\|K(X_T - y_0)\|] < \infty$. By Exercise 6.1.8, $\mathbb{E}_{x_0}[K^-(X_T - y_0)]$ is finite for all y. For any y,

$$\mathbb{E}_{x_0}[K(X_{T_n} - y_0)] = \mathbb{E}_{x_0}[K^+(X_T - y_0) : T_n = T]$$
$$- \mathbb{E}_{x_0}[K^-(X_T - y_0) : T_n = T]$$
$$+ \mathbb{E}_{x_0}[K(X_T - y_0) : T_n < T]. \quad (7.2.3)$$

$\mathbb{P}_{x_0}[T_n < T]$ tends to zero. On the set $T_n < T$, $|X_{T_n}| = n$ and $|X_{T_n} - y|/n$ tends boundedly to 1. Putting $y = y_0$ in (7.2.3) and taking limits shows (since $\mathbb{E}_{x_0}[\|K(X_T - y_0)\|] < \infty$ and $s(x_0, y_0) > -\infty$) that $\lim_{n \to \infty} (\log n) \mathbb{P}_{x_0}(T_n < T)$ exists and is finite. From (7.2.3), letting $n \to \infty$, for any y,

$$s(x_0, y) = \mathbb{E}_{x_0}[K^+(X_T - y)] - \mathbb{E}_{x_0}[K^-(X_T - y)]$$
$$- \lim_{n \to \infty} (\log n) \mathbb{P}_{x_0}(T_n < T).$$

The left side of the last equality can at worst be $-\infty$ while the right side at worst $+\infty$. One concludes that $s(x_0, y) > -\infty$ for all y and $\mathbb{E}_{x_0}[\|K(X_T - y_0)\|] < \infty$ for all y. s is symmetric. So $s(x, y) > -\infty$ for all x, provided $y \in D$.

Let ξ be arbitrary and $y_0 \in D$. From the last paragraph, $s(\xi, y_0) > -\infty$. As before (since $|X_{T_n} - y_0|$ is bounded away from zero, etc.), $\mathbb{E}_\xi[\|K(X_T - y_0)\|] < \infty$, $\lim_{n \to \infty} (\log n) \mathbb{P}_\xi(T_n < T)$ exists and is finite and $\mathbb{E}_\xi[K(X_T - \cdot)]$ is superharmonic. The last two facts imply, as before, that $s(\xi, y) > -\infty$ for all y, and

$$s(\xi, y) = \mathbb{E}_\xi[K(X_T - y)] - \lim_{n \to \infty} (\log n) \mathbb{P}_\xi(T_n < T).$$

Let us collect all the above in the following:

Proposition 7.2.1. *With the above notation, $\mathbb{E}_x[K(X_T - y)]$ is superharmonic in y for every $x \in \mathbb{R}^2$. And*

$$\begin{cases} s(x,y) = \mathbb{E}_x[K(X_T - y)] - a_D(x), & x, y \in \mathbb{R}^2 \\ a_D(x) = \lim_{n \to \infty} (\log n) \mathbb{P}_x(T_n < T), & x \in \mathbb{R}^2. \end{cases} \quad (7.2.4)$$

Remark. From (7.2.4), for each x, $s(x, \cdot)$ is superharmonic, and by symmetry, $s(\cdot, \cdot)$ is separately superharmonic. It is also easy to see that

$$\lim_{n \to \infty} \mathbb{E}_x[|K(X_{T_n} - y)|] = \mathbb{E}_x[|K(X_T - y)|] + a_D(x), \quad x, y \in \mathbb{R}^2. \quad (7.2.5)$$

Since $s(\cdot, \cdot)$ is separately harmonic in D, (7.2.4) allows that a_D is positive and harmonic in D.

Corollary 7.2.2. *For all $x \in \mathbb{R}^2$ and $b \in \partial D$,*

$$\liminf_{D \ni y \to b} \mathbb{E}_x[K(X_T - y)] = \mathbb{E}_x[K(X_T - b)], \quad x \in \mathbb{R}^2. \quad (7.2.6)$$

If $b \in \partial D$ is regular, the $\liminf$ in (7.2.6) can be replaced by a limit. If $b \notin \overline{D}$ or $b \in \partial D$ regular,

$$\mathbb{E}_x[K(X_T - y)] = K(x - b) + a_D(x), \quad x \in \mathbb{R}^2, \quad (7.2.7)$$

where a_D is defined in (7.2.4).

Proof. If $b \in \partial D$ is irregular, (7.2.6) follows from Exercise 6.1.9. Now suppose $b \in \partial D$ is regular. By lower semi-continuity, the left side in (7.2.6) is at last equal to the right side. Since $s(x, y) \leq K(x - y)$, we obtain by (7.2.4)

$$\liminf_{D \ni y \to b} \mathbb{E}_x[K(X_T - y)] \leq K(x - y) + a_D(x)$$

and the right side of the above inequality is just $\mathbb{E}_x[K(X_T - b)]$, as is seen by using (7.2.4) and recalling that $s(x, b) = K(x - b)$ if $b \in \partial D$ is regular. Equation (7.2.7) follows from (7.2.4). $\square$

Clearly, Corollary 7.2.2 says nothing if $x \notin \overline{D}$ or $x \in \partial D$ is regular. It is not difficult to show that a_D is non-negative harmonic in D and

$$\lim_{D \ni y \to b} a_D(x) = 0, \quad \text{if } b \in \partial D \text{ is regular.}$$

Green Functions, Boundary Value Problems, and Kernels

Theorem 7.2.3. *Let m be a probability measure on $\mathbb{R}^2$ such that*

$$s(x) = \int K(x-y)\, m(dy) \tag{7.2.8}$$

is superharmonic. Then $\mathbb{E}_x\left[s^-(X_T)\right] < \infty$ for all x and

$$\mathbb{E}_x\left[s(X_T)\right] \leq s(x) + a_D(x), \quad x \in \mathbb{R}^2. \tag{7.2.9}$$

If m does not charge D or the set of irregular points in ∂D, then equality obtains in (7.2.9).

Proof. We claim that

$$\int \mathbb{E}_x\left[K^-(X_T - y)\right] m(dy) < \infty, \quad x \in \mathbb{R}^2, \tag{7.2.10}$$

which of course implies that $\mathbb{E}_x\left[s^-(X_T)\right] < \infty$. For any y_0,

$$K^-(X_T - y_0) \leq K^-(2(X_T - y_0)) + K^-(2(y - y_0))$$
$$\leq \log 4 + K^-(X_T - y_0) + K^-(y - y_0).$$

Since $\mathbb{E}_x[K(X_T - \cdot)]$ and $\int K(\cdot - z)\, m(dz)$ are superharmonic,

$$\mathbb{E}_x\left[K^-(X_T - y)\right] < \infty \quad \text{and} \quad \int K^-(y_0 - z)\, m(dz) < \infty$$

for all y_0 (Exercise 6.1.8). Equation (7.2.10) is thus established and because of (7.2.10) use of Fubini is permitted in the following. Using (7.2.4) and that $s(x,y) \leq K(x-y$, we obtain

$$\mathbb{E}_x\left[s(X_T)\right] = \int \mathbb{E}_x\left[K(X_T - y)\right] m(dy)$$
$$\leq \int K(x-y)\, m(dz) + a_D(x) = s(x) + a_D(x).$$

Similarly, for the last statement, use (7.2.7). □

Example 7.2.4. In very simple cases, (7.2.7) can be used to find a_D. If D is the complement of the closed unit disc and $b = 0$, we get at once from (7.2.7) $a_D(x) = \log|x|$ for $|x| > 1$. If D is the upper half-plane, X_T lies on the real axis. For any b of the form $b = (0, -n)$ and z real, clearly $|z - b| \geq n$. It follows that $\mathbb{E}_x\left[K(X_T - y)\right] \leq \log \frac{1}{n}$. Using (7.2.7) and letting $n \to \infty$, we get $a_D(x) \leq 0$, i.e., $a_D \equiv 0$. Consider again the strip $\{0 < \Im z < 1\}$ for

$z \in \partial D$, $|z - n| \geq n - 1$ so that for $x \in D$, $\mathbb{E}_x [K(X_T - y)] \leq \log \frac{1}{n-1}$. Use (7.2.7) and let $n \to \infty$ to get $a_D = 0$. Similar arguments can be used to show that $a_D = 0$ for a wedge, a quadrant, etc. One may suspect that simple connectivity of a Green domain D is sufficient to guarantee that $a_D = 0$. This will turn out to be correct.

Again, in some simple cases, (7.2.7) can be used to compute the harmonic measure at x, i.e., the distribution of X_T relative to $\mathbb{P}_x$. As an example, consider $D =$ the upper half-plane. Denote the points on the plane by (x, y). If $b > 0$, the point $(a, -b) \notin \overline{D}$. From (7.2.7),

$$\int \log \left[(z - a)^2 + b^2 \right] \mathbb{P}_{(x,y)} (dz) = \log \left[(x - a)^2 + (y + b)^2 \right],$$

where $\mathbb{P}_{(x,y)} (dz) = \mathbb{P}_{(x,y)} (X_T \in dz)$. Differentiate both sides relative to b:

$$\int \frac{b}{(z - a)^2 + b^2} \mathbb{P}_{(x,y)} (dz) = \frac{y + b}{(x - a)^2 + (y + b)^2}.$$

Multiply both sides by $e^{i\alpha a}$ and integrate relative to da from $-\infty$ to ∞,

$$e^{-|\alpha|b} \hat{\mathbb{P}}_{(x,y)} (\alpha) = e^{i\alpha x} e^{-|\alpha|(y+b)},$$

where $\hat{\mathbb{P}}_{(x,y)} (\alpha) = \int e^{i\alpha z} \mathbb{P}_{(x,y)} (dz)$. Inversion shows that $\mathbb{P}_{(x,y)} (dz)$ has density

$$\frac{1}{\pi} \frac{y}{(x - z)^2 + y^2}.$$

We have thus found the Poisson kernel for the half-plane $\{(x, y) : y > 0\}$. (See Section 5.2.)

Example 7.2.5. As another example, consider the strip $\{(x, y) : 0 < x < 1\} = D$. Fix $(x, y) \in D$. $\mathbb{P}_{(x,y)} (X_T \in dz)$ lives on the lines $x = 0$ and $x = 1$. Denote these parts by Q_0 and Q_1, respectively. If $a > 1$, the point $(a, b) \notin \overline{D}$. From (7.2.7),

$$\int \log \left[a^2 + (b - z)^2 \right] Q_0 (dz) + \int \log \left[(a - 1)^2 + (b - z)^2 \right] Q_1 (dz)$$
$$= \log \left[(x - a)^2 + (y - b)^2 \right].$$

As before, differentiating relative to a then multiplying both sides by $e^{i\alpha b}$ and integrating relative to db from $-\infty$ to ∞:

$$\hat{Q}_0 (\alpha) + e^{|\alpha|} \hat{Q}_1 (\alpha) = e^{|\alpha|x} e^{i\alpha y},$$

where $\hat{Q}_j (\alpha) = e^{i\alpha z} Q_j (dz)$, $j = 0, 1$.

Also the point $(-a, b) \notin \overline{D}$ if $a > 0$. Repetition of the above leads to
$$\hat{Q}_0(\alpha) + e^{-|\alpha|}\hat{Q}_1(\alpha) = e^{-|\alpha|x}e^{i\alpha y}.$$
We must thus have
$$e^{|\alpha|}\left[1 - e^{-2|\alpha|}\right]\hat{Q}_1(\alpha) = e^{i\alpha y}\left[e^{|\alpha|x} - e^{-|\alpha|x}\right]$$
which expands into
$$\hat{Q}_1(\alpha) = \sum_0^\infty e^{i\alpha y}\left\{e^{-(2n+1-x)|\alpha|} - e^{-(2n+1+x)|\alpha|}\right\}.$$
$Q_1(dz)$ thus has density
$$\hat{Q}_1(\alpha) = \frac{1}{\pi}\sum_0^\infty \left\{\frac{2n+1-x}{(2n+1-x)^2 + (y-z)^2} - \frac{2n+1+x}{(2n+1+x)^2 + (y-z)^2}\right\},$$
and $Q_0(dz)$ has density
$$\frac{1}{\pi}\sum_0^\infty \left\{\frac{2n+x}{(2n+x)^2 + (y-z)^2} - \frac{2n+2-x}{(2n+2-x)^2 + (y-z)^2}\right\}.$$
The above can be interpreted in the "image method". See remark after Example 7.4.2.

Exercises

7.2.1. Show that
$$\lim_{D \ni x \to b} a_D(x) = 0 \quad \text{if } b \in \partial D \text{ is regular}.$$
Hint: Let $y_0 \in D$. Then,
$$G_D(x, y_0) = K(x - y_0) - \mathbb{E}_x\left[K(X_T - y_0)\right] + a_D(x).$$
As $x \to b$, $\lim G_D(x, y_0) = 0$. Thus, as $D \ni x \to b$,
$$\limsup a_D(x) + \liminf \{K(x - y_0) - \mathbb{E}_x[K(X_T - y_0)]\} \leq 0.$$
If $m = K(b - y_0)$, $K(\cdot - y_0) \vee m$ is bounded and continuous on ∂D. So
$$\limsup_{x \to b} \mathbb{E}_x\left[K(X_T - y_0)\right] \leq \limsup_{x \to b} \mathbb{E}_x\left[K(X_T - y_0) \vee m\right]$$
$$= K(b - y_0) \vee m = K(b - y_0).$$

7.2.2. Show that
$$a_D(x) \leq \log|x| + O(1).$$

Hint: Let $y_0 \in D$. Then
$$G_D(x, y_0) = K(x - y_0) - \mathbb{E}_x[K(X_T - y_0)] + a_D(x)$$
is bounded off any neighborhood of y_0. Also since $|X_T - y_0|$ is bounded away from zero, $\mathbb{E}_x[K(X_T - y_0)]$ is bounded above. Finally, a_D being harmonic in D is bounded in a neighborhood of y_0.

7.2.3. Show that
$$\limsup_{x \to \infty} \frac{a_D(x)}{\log|x|} = 1$$
unless $a_D \equiv 0$.

Hint: a_D is harmonic in D and bounded in D_n (by the above exercise). Hence,
$$a_D(x) = \mathbb{E}_x[a_D(X_T) : T_n < T].$$

Therefore,
$$a_D(x) \leq \left(\sup_{|x|=n} \frac{a_D(x)}{\log n}\right)(\log n \, \mathbb{P}_x[T_n < T]),$$

i.e.,
$$a_D(x) \leq \left(\limsup_{|x| \to \infty} \frac{a_D(x)}{\log|x|}\right) a_D(x).$$

7.3 Unbounded Open Sets (*Continued*)

Let D be an open set having a Green function. Retaining the notion of Section 7.2, we have the following expression for G_D:
$$G_D(x, y) = K(x - y) - \mathbb{E}_x[K(X_T - y)] + a_D(x), \quad x, y \in D. \quad (7.3.1)$$
Thus, G_D is given by (7.1.5) if and only if $a_D \equiv 0$.

Proposition 7.3.1. *For any probability measure m on $\mathbb{R}^2$ satisfying*
$$\int \log(1 + |y|)\, m(dy) < \infty, \quad (7.3.2)$$
we have
$$\limsup_{x \to \infty} \left\{\int \log|x - y|\, m(dy) - \log|x|\right\} = 0. \quad (7.3.3)$$

Proof. If $f(x)$ is the function in the brackets in (7.3.3),

$$f\left(\frac{1}{x}\right) = \int \log|1 - xy|\, m\,(dy), \quad x \neq 0, \tag{7.3.4}$$

and the right side of (7.3.4) is subharmonic on $\mathbb{R}^2$, under (7.3.2). (Indeed, $\log|1 - xy|$ is subharmonic in x because $1 - yx$ is analytic in x. If $|x| \leq n$, $\log|1 - xy|$ is bounded above by the m-integrable function $n\log(1 + |y|)$, Fatou and Fubini are therefore applicable.) The lim sup in (7.3.3) is thus the value of this subharmonic function at 0, namely 0.

Remark. Note that (7.3.2) is necessary and sufficient that

$$\int \log|x - y|\, m\,(dy)$$

be subharmonic on $\mathbb{R}^2$.

The following theorem gives necessary and sufficient conditions that $a_D \equiv 0$, i.e., (7.1.5) is valid.

Theorem 7.3.2. *Let D be Greenian. Then*

$$a_D(x) = \limsup_{D \ni y \to \infty} G_D(x, y), \quad x \in D. \tag{7.3.5}$$

Proof. From the above remark and Proposition 7.3.1, for any $x \in D$,

$$\limsup_{y \to \infty} \{K(x - y) - \mathbb{E}_x[K(X_T - y)]\} = 0. \tag{7.3.6}$$

If $a_D(x) = 0$, we see from (7.3.1) that

$$0 \leq \limsup_{D \ni y \to \infty} G_D(x, y) = \limsup_{D \ni y \to \infty} \{K(x - y) - \mathbb{E}_x[K(X_T - y)]\} \leq 0.$$

The last inequality follows from (7.3.6).

Now suppose $a_D(x) > 0$. If $y \notin \overline{D}$, by (7.2.7), the quantity in brackets of (7.3.6) is just $-a_D(x)$. Also (7.2.6) says that for any $b \in \partial D$, $\mathbb{E}_x[K(X_T - b)]$ is a limit point of $\mathbb{E}_x[K(X_T - y)]$ as $D \ni y$ tends to b. Thus, the lim sup in (7.3.6) can be taken as y tends to infinity in D. Taking lim sup in (7.3.1) gives us (7.3.5). □

The following proposition gives a simple geometrical condition for regularity and can also be proved using the fact that the Brownian path winds around its starting point infinitely often. See McKean Jr. [8].

Proposition 7.3.3. *Let D be an open set. If $b \in \partial D$ is contained in a continuum completely contained in the complement of D, then b is regular.*

Proof. Let $b \in F \subset \mathbb{R}^2 \backslash D$ be a continuum. If r is small, $F \cap \{x : |x - b| \geq r\} \neq \emptyset$. Let D_1 be the open set

$$D_1 = \{x : |x - b| < r\} \backslash F.$$

D_1 is bounded open, $D_1 \supset D \cap \{x : |x - b| < r\}$, $b \in \partial D$, and ∂D_1 is connected. D_1 is simply connected because ∂D_1 is connected. By the Riemann mapping theorem, there is a 1-1 holomorphic map f defined on D_1 such that $f(D_1)$ is the open unit disc. It is easily seen that as $x \in D_1$ tends to any point in ∂D_1, $|f(x)|$ tends to 1. The function $h(x) = 1 - |f(x)|$ is thus a barrier at all points of ∂D_1. Every point in ∂D_1, in particular b, is thus regular for D_1^c. Since $b \in \partial D$ and $D_1 \supset D \cap \{x : |x - b| < r\}$, b is regular for $\{D \cap \{x : |x - b| < r\}\}^c$ and hence for D^c. That proves the proposition. $\square$

Remark. Let D be an unbounded Green domain. Assume $0 \notin D$ and consider the map $x \mapsto \frac{1}{x}$. If D_1 is the image of D under this map, $0 \in \partial D_1$, and $G_{D_1}(x, y) = G_D\left(\frac{1}{x}, \frac{1}{y}\right)$. $\lim_{D \ni y \to \infty} G_D(x, y) = 0$ is equivalent to $\lim_{D_1 \ni y \to \infty} G_{D_1}(x, y) = 0$, i.e., that (Property 2, Section 7.1) 0 is regular for the complement of D_1. Proposition 7.3.3 thus implies the following: If D is an unbounded domain such that $\mathbb{R}^2 \backslash D$ has an unbounded component, then $a_D \equiv 0$. In particular, if $D \neq \mathbb{R}^2$ is simply connected, then $a_D \equiv 0$.

The case of D with compact complement. When the complement of D is compact, a_D can be expressed in a nice way as follows. From (7.2.4), and the symmetry of s,

$$a_D(x) = \{\mathbb{E}_x[K(X_T - y) + a_D(y)]\} - \mathbb{E}_y[K(X_T - x)], \quad x, y \in \mathbb{R}^2. \tag{7.3.7}$$

Let B be a disc containing the complement of D and let $S = $ hitting time to B. Then

$$\mathbb{E}_y[K(X_T - x)] = \mathbb{E}_y[\mathbb{E}_{X_S}[K(X_T - x)]]. \tag{7.3.8}$$

For a sequence y_n tending to infinity, the probability measures $\mathbb{P}_{y_n}[X_S \in dz]$ converge to a probability measure μ, say, on ∂B. Since $\mathbb{E}.[K(X_T - x)]$ is continuous on ∂B, as y_n tends to infinity, the left side of (7.3.8) tends to

$$\int \mathbb{E}_z[K(X_T - x)]\mu(dz) = \int K(z - x)m(dz), \tag{7.3.9}$$

where

$$m(dz) = \int \mathbb{P}_y[X_T \in dz]\,\mu(dz). \tag{7.3.10}$$

It follows that the term in brackets in (7.3.7) converges as $y_n \to \infty$, and, because $\log \frac{|z-y_n|}{|y_n|}$ tends to zero uniformly for $z \in \partial D$, this limit is simply

$$\lim_{y_n \to \infty} [a_D(y_n) - \log|y_n|] = r, \text{ say.} \tag{7.3.11}$$

Thus we obtain from (7.3.7)

$$a_D(x) = r - \int K(x-z)\,m(dz), \quad x \in \mathbb{R}^2, \tag{7.3.12}$$

where m is a probability measure on ∂D as is clear from (7.3.10).

Remark. Given a compact set C, the Robin problem is to find a measure m on C whose logarithmic potential, i.e., $\int K(\cdot - z)$, is constant on $C \setminus$ (irregular points). If $C =$ the complement of D, we see from (7.3.12) that m solves the Robin problem for C.

The third term in (7.3.12), being superharmonic on $\mathbb{R}^2$, is bounded below on compact sets. It is also clear from (7.3.12) that

$$\lim_{x \to \infty} (a_D(x) - \log|x|) = r. \tag{7.3.13}$$

r is called the Robin constant for D.

Poisson equation. If D is Greenian and f is C^1 on $\mathbb{R}^2$ with bounded support,

$$u(x) = \int G_D(x,y)\,f(y)\,dy, \quad x \in D,$$

is obvious C^2 on D and satisfies $\Delta u = -2\pi f$. This is so because $G_D - K$ is harmonic in D. Let us show that u is bounded in D and $u(x)$ tends to zero as $x \to b \in \partial D$ regular. This we do by reducing to the case where ∂D is compact as follows: Let $b \in \partial D$ be regular and $x_0 \in D$. Let A be a compact subset of ∂D, containing b such that $\mathbb{P}_{x_0}[X_T \in A] > 0$. If B is a disc containing A and $W = D \cup B^c$, then W^c is compact, $b \in \partial W$, and is regular for $(B \cap D)^c = (B \cap W)^c$ and regularity is a local property. In particular, by Theorem 6.1.10, W is Greenian. Since $W \supset D$, $G_W \geq G_D$.

Thus, it is sufficient to show the following: Let f be bounded measurable with bounded support and D Greenian with D^c compact. Then

$$u(x) = \int G_D(x,y) f(y) \, dy \qquad (7.3.14)$$

is bound in D and $u(x) \to 0$ as $x \to b \in \partial D$ regular. If $F(y) = \int K(y-z) f(z) dz$, u of (7.3.14) has the form

$$u(x) = F(x) - \mathbb{E}_x[F(X_T)] + a_D(x) \int f(y) \, dy. \qquad (7.3.15)$$

F is continuous on $\mathbb{R}^2$ so F is bounded and continuous on ∂D. Therefore, the second term in (7.3.15) tends to $F(b)$ as $x \to b \in \partial D$ regular. And $a_D(x)$ tends to zero as $x \to b$ (Exercise 7.2.1).

Finally, since $F(x) = \log(x) \int f(y) \, dy + O(1)$, boundedness of u is clear from (7.3.13).

Exercises

7.3.1. Let D have a Green function. For any bounded subset A of D and any $\epsilon > 0$, show that

$$\sup G_D(x,y) < \infty,$$

where the sup is over all $x \in A$ and $y \in D$ such that $|x - y| \geq \epsilon$.

Hint: Reduce to the case where D^c is compact.

7.3.2. In the notation of Section 7.2, show that $\mathbb{P}_x[T_n < T]$ tends to zero uniformly on compact subsets of $\mathbb{R}^2$.

Remark. If we reduce to the case where D^c is compact, this is clear from (7.3.13) and the expression of a_D given in (7.2.4). But show this directly.

7.3.3. In the notation of Section 7.2, show that $s_n(\cdot, \cdot)$ decreases uniformly on bounded subsets of $\mathbb{R}^2 \times \mathbb{R}^2$ to $s(\cdot, \cdot)$.

Hint: For $n < m$,

$$s_n(x,y) = s_m(x,y) + \mathbb{E}_x[G_{D_m}(X_{T_n}, y) : T_n < T].$$

Note that $G_{D_m} \leq G_D$. Now use Exercises 7.3.1 and 7.3.2.

Remark. This provides another proof that $s(\cdot, \cdot)$ is separately superharmonic and lower semicontinuous in both variables.

7.4 Examples

Example 7.4.1. $\mathbb{R}^1$ does not have a Green function but any open set $D \subset \mathbb{R}^1$, whose complement is not empty, has a Green function. In fact, $u(x) = |x|$ is positive and harmonic in the open set $\mathbb{R}^1 \setminus \{0\}$.

The Green function for a finite open interval (a, b) is

$$G(x, y) = -|x - y| + \frac{(a+b)(x+y) - 2(xy+ab)}{b-a} \qquad (7.4.1)$$

and of a half-line (a, ∞) is

$$G(x, y) = -|x - y| - 2a + (x + y). \qquad (7.4.2)$$

Indeed (by looking at the harmonic function with boundary value $= 1$ at a, and $= 0$ at b), it is seen that $\mathbb{P}_x(X_T = a) = \frac{b-x}{b-a}$. Thus, the Green function of (a, b) is

$$G(x, y) = -|x - y| + \mathbb{E}_x[|X_T - y|]$$
$$= -|x - y| + (y - a)\mathbb{P}_x(X_T = a) + (b - y)\mathbb{P}_x(X_T = b)$$

which is the expression given in (7.4.1). Letting $b \to \infty$ in (7.4.1), one gets the Green function for (a, ∞).

Equation (7.4.2) resembles (7.3.1), with $a_D(x) = x - a$ which is positive harmonic in (a, ∞). It follows that formula (7.1.5) holds for an unbounded open set in $\mathbb{R}^1$ if and only if it has no unbounded components, i.e., if and only if it is the disjoint union of bounded open intervals.

Example 7.4.2. The Green function of the ball $D = B(0,1)$ in $\mathbb{R}^d$ with center 0 and radius r is

$$G_D(x, y) = \log\left|\frac{r^2 - x\bar{y}}{r(x-y)}\right|, \quad |x|, |y| < r, \ d = 2,$$

$$G_D(x, y) = \frac{1}{|x-y|^{d-2}} - \frac{r^{d-2}}{|y|^{d-2}} \frac{1}{|x-y^*|^{d-2}}, \quad d \geq 3, \qquad (7.4.3)$$

where $\bar{y}$ is the conjugate of y and y^* is the inverse of y relative to $\partial B(0, r)$: $y^* = \frac{r^2}{|y|^2} y$.

To show (7.4.3), we evaluate $\mathbb{E}_x[K(X_T - y)]$, $T = $ exit time from D. For $z \in \partial B(0, r)$,

$$\frac{|z - y^*|}{|z - y|} = \frac{r}{|y|}.$$

So

$$K(X_T - y) = \begin{cases} \log \frac{r}{|r|} + K(X_T - y^*), & d = 2, \\ \frac{r^{d-2}}{|y|^{d-2}} K(X_T - y^*), & d \geq 3. \end{cases}$$

$K(\cdot - y^*)$ is continuous on $\overline{D}$ and harmonic in D giving

$$\mathbb{E}_x [K(X_T - y)] = \begin{cases} \log \frac{r}{|r|} + K(X_T - y^*), & d = 2, \\ \frac{r^{d-2}}{|y|^{d-2}} K(X_T - y^*), & d \geq 3. \end{cases}$$

Remark. The method in Example 7.4.2 is sometimes referred to as the "method of images" due to Lord Kelvin. In this method, one tries to guess, using the geometry of the domain, at a suitable "distribution of charges" outside the domain whose potential on the boundary equals that of a point charge inside the domain; this means that we try to guess at a function which is harmonic in D and whose values are equal to $K(x - \cdot)$ on ∂D. Its applicability is limited; nonetheless, it is very ingenious. When it is applicable, it allows us to find the Green function almost without computation. Let us give a few examples to make the idea a little clearer.

Example 7.4.3. The Green function of the half-space

$$D = \{x = (x_1, \ldots, x_d) : x_d > 0\}$$

is

$$G_D(x, y) = K(x - y) - K(x - y^*), \tag{7.4.4}$$

where y^* is the image of y in the plane ∂D: If $y = (y_1, \ldots, y_d)$,

$$y^* = (y_1, \ldots, y_{d-1},, -y_d).$$

Clearly, for $z \in \partial D$, $|y - z| = |y^* - z|$ so that $K(X_T - y) = K(X_T - y^*)$, where T = exit time from D. $K(\cdot - y^*)$ is harmonic in D and continuous on $\overline{D}$. If $d \geq 3$, further $K(z - y^*)$ tends to zero as $z \to \infty$. We must have

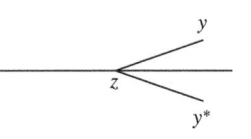

$$K(z - y^*) = \mathbb{E}_x [K(X_T - y)] \tag{7.4.5}$$

giving (7.4.4) for $d \geq 3$. When $d = 2$, $K(z - y^*)$ is not bounded and we cannot at once claim (7.4.5). The results in Section 7.3, however, guarantee that (7.4.4) and (7.4.5) hold for $d = 2$ as well.

Example 7.4.4. Let us use the method of images to find the Green function of a quadrant. For simplicity, we take the case $d = 2$; the method is almost exactly the same for $d \geq 3$. Let D denote the quadrant

$$D = \{z : \Re z > 0,\ \Im z > 0\}.$$

We know from Section 7.3 that (7.1.5) is valid in this case, and the problem is to find $\mathbb{E}_x \left[K \left(X_T - y \right) \right]$.

If $y_1 \,(= \bar{y})$ is the image of y in $(\Im z = 0)$, y_2 the image of y_1 in $(\Re z = 0)$, and y_3 the image of y_2 in $(\Im z = 0)$, then

$$K(z - y_1) - K(z - y_2) + K(z - y_3) = K(z - y)$$

for all $z \in \partial D$ as is easily seen. See Figure 7.1. From Section 7.3 follows

$$\mathbb{E}_x \left[K \left(X_T - y \right) \right] = K(x - y_1) - K(x - y_2) + K(x - y_3).$$

Since $y_1 = \bar{y}$, $y_2 = -y$, $y_3 = -\bar{y}$, we find that the Green function of D is

$$\log \left| \frac{x^2 - \bar{y}^2}{x^2 - y^2} \right|.$$

The above argument extends to any wedge with an angle $2\pi/n$.

For $y \in D$ (see Figure 7.2), let y_1 be the image of y in L_1, y_2 the image of y_1 in the line L_2, y_3 the image of y_2 in the line L_1, etc. The Green function of D is then

$$K(x - y) - K(x - y_1) + K(x - y_2) - K(x - y_3) + \text{etc.}$$

The sum has only n terms.

Fig. 7.1 Green function in a quadrant.

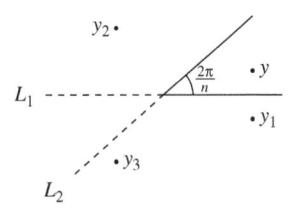

Fig. 7.2 Green function for a wedge.

Example 7.4.5. The method of images can be used to find the Green function of the region between two parallel hyperplanes. One needs an infinite series of reflections and has to watch out when summing the resulting series. The Green function of the strip $D = \{0 < \Im z < \pi\}$ can be computed to be

$$\sum_{-\infty}^{\infty} \{K(x - y - 2ni\pi) - K(x - \bar{y} + 2ni\pi)\} = \sum_{-\infty}^{\infty} \log \left| \frac{x - \bar{y} + 2ni\pi}{x - \bar{y} - 2ni\pi} \right|. \tag{7.4.6}$$

In Courant–Hilbert [2, p. 378], the Green function of a rectangular parallelepiped is computed using the image method and is given by an infinite series which has to be grouped correctly for convergence.

Let us give one last example of the image method.

Example 7.4.6. Let D be the spherical shell

$$K = \{x : r < |x| < R\}$$

in $\mathbb{R}^d$, $d \geq 3$. Let $y = y_0 \in D$ and define y_n, $n = 1, 2, \ldots$, inductively as follows: y_1 is the image of y_0 in the sphere $\{x : |x| = R\}$, i.e., $y_1 = y^* = R^2 \frac{y}{|y|^2}$. y_2 is the image of y_1 in the sphere $\{x : |x| = r\}$, y_3 is the image of y_2 in the sphere $\{x : |x| = R\}$, etc. Similarly, define y_{-n}, $n = 1, 2, \ldots$, as follows: y_{-1} is the image of y_0 in $\{x : |x| = r\}$, y_{-2} the image of y_{-1} in $\{x : |x| = R\}$, etc. If $a = \frac{R}{r}$, it is seen that

$$y_{2n} = a^{-2n} y,$$

$$y_{2n+1} = a^{2n} y^*, \quad y^* = \frac{R^2 y}{|y|^2},$$

$n = 0, \pm 1, \pm 2, \ldots$. And the Green function of D is

$$\sum_{-\infty}^{\infty} a^{-n(d-2)} K(x - y_{2n}) - \frac{R^{d-2}}{|y|^{d-2}} \sum_{-\infty}^{\infty} a^{n(d-2)} K(x - y_{2n+1}).$$

The series above converges uniformly and absolutely in $\overline{D}$. See also the remark after Example 7.4.8.

Example 7.4.7. For plane regions, conformal maps provide a very powerful method of determining Green functions. The Riemann mapping theorem in theory permits us to write down the Green function of any simply connected domain with at least two boundary points.

The function $w(z) = \frac{1+iz}{z+i}$ maps the upper half-plane $\{\Im z > 0\}$ 1-1 conformally onto the unit disc. By Example 7.4.2, the Green function of the upper half-plane is

$$\log \left| \frac{x - \bar{y}}{x - y} \right|, \quad \Im x > 0, \ \Im y > 0. \tag{7.4.7}$$

The function $w(z) = e^z$ maps the strip $\{0 < \Im z < \pi\}$ 1-1 conformally onto the upper half-plane. From (7.4.7), the Green function of the strip $\{0 < \Im z < \pi\}$ is

$$\log \left| \frac{e^{x-\bar{y}} - 1}{e^{x-y} - 1} \right|. \tag{7.4.8}$$

We leave it as an exercise to show that the expressions in (7.4.6) and (7.4.8) represent the same function; consult Konrad Knopp [16, p. 82], Exercise 11a, if needed.

Again the function $w(z) = e^z$ maps the strip $\{0 < \Im z < 2\pi\}$ conformally onto the plane cut along the positive real axis. Using (7.4.8), the Green function of the plane slit along the positive real axis is found to be

$$\log \left| \frac{\sqrt{x} - \sqrt{\bar{y}}}{\sqrt{x} - \sqrt{y}} \right|.$$

In the following example, we use the method of separation of variables.

Example 7.4.8. Let us compute the Green function for the circular ring

$$D = \{x \in \mathbb{R}^2 : a < |x| < b\}.$$

The geometry of D suggests use of polar coordinates. In polar coordinates, the Laplacian has the form

$$\frac{\partial^2}{\partial r^2} + \frac{1}{r} \frac{\partial}{\partial r} + \frac{1}{r^2} \frac{\partial^2}{\partial \theta^2}. \tag{7.4.9}$$

An attempt at finding a harmonic function of the form $f(r)g(\theta)$ leads us to the equations

$$f'' + \frac{1}{r} f' + \frac{\lambda}{r^2} f = 0,$$

$$\frac{d^2 g}{d\theta^2} = \lambda g,$$

where λ is a constant. Since g must be periodic with period 2π, the possible values of λ (the eigenvalues) are $-n^2$, $n = 1, 2, \ldots$, and the corresponding independent solutions are, for $n \neq 0$, $e^{in\theta}$ and $e^{-in\theta}$.

Now, let us attempt to solve the Poisson equation

$$\frac{\partial^2 u}{\partial r^2} + \frac{1}{r}\frac{\partial u}{\partial r} + \frac{1}{r^2}\frac{\partial^2 u}{\partial \theta^2} = -f(r, \theta)$$

in D with boundary condition $u(a, \theta) = u(b, \theta) = 0$. Writing

$$u(r, \theta) = \sum u_n(r) e^{in\theta},$$

$$f(r, \theta) = \sum f_n(r) e^{in\theta},$$

we arrive at

$$u_n'' + \frac{1}{r} u_n' - \frac{n^2}{r^2} u_n = -f_n. \qquad (7.4.10)$$

The unique solution of (7.4.10) satisfying $u_n(a) = u_n(b) = 0$ is given by

$$u_n(s) = \int_a^b g_n(r, s) f_n(r) r\, dr,$$

where if $n \neq 0$

$$g_n(r, s) = \begin{cases} c_n \left(\frac{b^{2n}}{s^n} - s^n\right)\left(r^n - \frac{a^{2n}}{r^n}\right), & a \leq r \leq s \\ c_n \left(s^n - \frac{a^{2n}}{s^n}\right)\left(\frac{b^{2n}}{r^n} - r^n\right), & s \leq r \leq b \end{cases} \qquad (7.4.11)$$

with $c_n^{-1} = 2n(b^{2n} - a^{2n})$ and

$$g_0(r, s) = \begin{cases} c_0 \log \frac{b}{s} \log \frac{r}{a}, & a \leq r \leq s \\ c_0 \log \frac{s}{a} \log \frac{b}{r}, & s \leq r \leq b \end{cases} \qquad (7.4.12)$$

with $c_0^{-1} = \log \frac{b}{a}$.

Using $f_n(r) = \frac{1}{2\pi} \int_0^{2\pi} e^{-in\theta} f(r, \theta)\, d\theta$, we get an expression for u:

$$u(s, \theta) = \frac{1}{2\pi} \sum_{-\infty}^{\infty} \int_a^b dr \int_0^{2\pi} d\varphi\, g_n(r, s) e^{in(n-\varphi)} r f(r, \varphi). \qquad (7.4.13)$$

$r\, dr\, d\varphi$ being the area element in polar coordinates, we obtain from (7.4.13) a tentative expression for the Green function for D:

$$\sum_{-\infty}^{\infty} g_n(r, s) e^{in(\theta-\varphi)}, \qquad (7.4.14)$$

where $x = (r, \theta)$ and $y = (s, \varphi)$ and g_n are defined in (7.4.11) and (7.4.12). Since $g_n = g_{-n}$, (7.4.14) is in fact

$$g_0(r, s) + 2 \sum_1^\infty g_n(r, s) \cos n(\theta - \varphi). \tag{7.4.15}$$

We have yet to show that (7.4.15) in fact represents G_D. It can be verified that $\sum_1^\infty |g_n - g_{n+1}|$ converges uniformly in $a \leq r, s \leq b$. By Theorem (2, 6), of Zygmund [15, p. 4], (7.4.15) represents a continuous function of (x, y) if $\theta \neq \varphi$. On the other hand, if $r \neq s$ (i.e., $|x| \neq |y|$), $\sum g_n(r, s)$ is dominated by a geometric series ($g_n(r, s)$ is less or equal to a constant limes $\left(\frac{r}{s}\right)^n$ for $r \leq s$) and so (7.4.15) represents a continuous function of (x, y) if $|x| \neq |y|$. Together, these imply that (7.4.15) represents a continuous function of (x, y) outside of the diagonal $\{x = y\}$.

It can be checked that $\int_a^b g_n(r, s)\, dr = O\left(\frac{1}{n^2}\right)$. Using this and denoting the continuous function given in (7.4.15) by $F(x, y)$, we see that for a C^1-function h of the form $h(r, \theta) = f(r)e^{in\theta}$, the function

$$u(r, \theta) = \int_D F(x, y) h(y)\, dy = e^{in\theta} \int_a^b g_n(r, s) f(s)\, s\, ds$$

satisfies $\Delta u = -2\pi h$ and u vanishes on ∂D. We must have

$$u(x) = \int_D G_D(x, y) h(y)\, dy.$$

It follows that $F(x, y) = G_D(x, y)$ for almost all y and by continuity $F \equiv G_D$.

We have shown that (7.4.15) represents the Green function for the circular ring $\{x : a < |x| < b\}$.

Remark. Using Example 7.4.8 and inversion, one can determine the Green function for the region between two circles, one completely contained in the other; this is the shaded region in the following figure (Figure 7.3). See Exercise 5.3.2.

The method of separation of variables can be used whenever we have a product domain. For a more detailed account of this method, consult any elementary book on partial differential equations; see Courant–Hilbert [2] for a complex function method of determining the Green function of a circular ring.

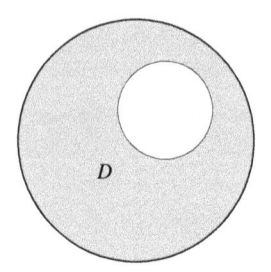

Fig. 7.3 Green function in the region between two circles.

7.5 The Green Function and Relative Transition

Let $W \subset \mathbb{R}^d$ be open and $T =$ exit time from W. The *relative transition measure* $Q(t, x, A)$ is

$$Q(t, x, A) = \mathbb{P}_x[X_T \in A, \ t < T], \quad A \text{ Borel}. \tag{7.5.1}$$

It is clear that $Q(t, x, \cdot)$ is absolutely continuous relative to Lebesgue measure. We shall find nice densities for Q and relate these to the Green function of W. $p(t, x, y)$ will denote the Gauss kernel:

$$p(t, x, y) = (2\pi t)^{-2/d} \exp\left(-\frac{|x-y|^2}{2t}\right), \quad t > 0.$$

We start with the following:

Proposition 7.5.1. *For all x and $t > 0$,*

$$\mathbb{P}_x(T = t) = 0. \tag{7.5.2}$$

Proof. Suppose (7.5.2) is false. On the set $T > s$, $T = s + T(\theta_s)$. By Markov property, we obtain for each $0 < s < t$,

$$\mathbb{E}_x\left[\mathbb{P}_{X_s}(T = t - s) : T > s\right] > 0.$$

The measures $\mathbb{P}_x(X_s \in \cdot)$ and $\mathbb{P}_x(X_1 \in \cdot)$ are clearly equivalent so that

$$\mathbb{E}_x\left[\mathbb{P}_{X_1}(T = t - s)\right] > 0$$

for $s < t$. This is impossible because $\mathbb{E}_x[\mathbb{P}_{X_1}(T \in \cdot)]$ is a finite measure. □

By the first passage time relation (3.2.8),

$$\mathbb{P}_x(X_t \in A) = Q(t, x, A) + \int_{[0,t] \times \partial W} \mathbb{P}_b(X_{t-s} \in A) \mathbb{P}_x(T \in ds, X_t \in db). \tag{7.5.3}$$

If $q(t,x,y)$ denotes the density of $Q(t,x,\cdot)$, from (7.5.3) for almost all y,

$$p(t,x,y) = q(t,x,y) + \mathbb{E}_x\left[q(t-T, X_T, y) : T < t\right]. \qquad (7.5.4)$$

In the last expectation, using (7.5.2), we have replaced $(T \le t)$ by $(T < t)$. Now $p(\cdot,\cdot,\cdot)$ is lower semicontinuous, being the increasing limit of continuous functions

$$(2\pi t)^{-2/d} \exp\left(-\frac{|x-y|^2 + \epsilon}{2t}\right).$$

Therefore, the last term in (7.5.4) is lower semicontinuous by Fatou. q defined by (7.5.4) is therefore upper semicontinuous in $\{(t,x,y) : t > 0\}$. For fixed t, x, it is non-negative almost everywhere (being a density); hence it is non-negative everywhere. *We define q by (7.5.4) for all $t > 0$, $x, y \in \mathbb{R}^d$.*

Now p is C^∞ in $(0,\infty) \times \mathbb{R}^d \times \mathbb{R}^d$. Since $X_T \in \partial W$ for each x, we find from (7.5.4)

$q(t,x,y)$ is C^∞ in (t,y) if $y \notin \partial W$.

Using the semigroup property of $p(t,\cdot,\cdot)$, we see from (7.5.4)

$$p(t,x,y) = \int q(t-\epsilon, x, z) p(\epsilon, z, y) \, dz + \mathbb{E}_x\left[p(t-T, X_T, y) : T < t - \epsilon\right].$$

Comparing the above with (7.5.4),

$$\int q(t-\epsilon, x, z) p(\epsilon, z, y) \, dz = q(t,x,y) + \mathbb{E}_x\left[p(t-T, X_T, y) : t - \epsilon \le T < t\right]. \qquad (7.5.5)$$

In particular,

$$\lim_{\epsilon \to 0} \int q(t-\epsilon, x, z) p(\epsilon, z, y) \, dz = q(t,x,y), \quad t > 0, \ x, y \in \mathbb{R}^d. \qquad (7.5.6)$$

By Markov property, $Q(t,x,\cdot)$ is a semi-group of measures. In terms of densities, this means the following: For each x and $t, s > 0$, for almost all y,

$$q(t+s, x, y) = \int q(t, x, z) q(s, z, y) \, dz. \qquad (7.5.7)$$

But because of (7.5.6), it is immediately seen that (7.5.7) *holds for all y*. Equation (7.5.7) can be rewritten as

$$q(t+s, x, y) = \mathbb{E}_x\left[q(s, X_t, y) : t < T\right]$$

which shows that $q(\theta, x, y)$ is continuous in x for $x \in W$. ($q(u,\cdot,\cdot) \le p(u,\cdot,\cdot) \le (2\pi u)^{-d/2}$. Now refer to Exercise 5.2.3.) Using this and

Proposition 7.5.1, we see further (since $X_t \in W$ for $t < T$) that

$$\text{for all } x, y, \ q(\cdot, x, y) \text{ is continuous in } (0, \infty). \tag{7.5.8}$$

Now we shall relate these *relative transition densities* q to the Green function of W as follows. We assume from now on that $d = 2$. As the reader will see, the case $d \geq 3$ is similar and simpler.

Let $0 \leq f$ be C^1 on $\mathbb{R}^2$ with compact support, $D_n = W \cap \{0 \leq |x| < n\}$, and T_n = exit time from D. If $u = \int K(\cdot - y) f(y) \, dy$, then u is C^2 on $\mathbb{R}^2$ and $\Delta u = -2\pi f$ by Theorem 6.3.3. By Dynkin's formula, Theorem 5.1.2,

$$u(x) = \mathbb{E}_x \left[u(X_{T_n}) \right] + \pi \mathbb{E}_x \left[\int_0^{T_n} f(X_t) \, dt \right]. \tag{7.5.9}$$

Suppose now that W has a Green function. Let n tend to infinity in (7.5.9). With the notation of Section 7.2,

$$\int K(x, y) f(y) \, dy = \pi \mathbb{E}_x \left[\int_0^T f(X_t) \, dt \right] + \int s(x, y) f(y) \, dy \tag{7.5.10}$$

because $s(x, \cdot)$ being superharmonic on $\mathbb{R}^2$ is bounded below on compact sets. Deduction of (7.5.10) from (7.5.9) is thus justified. Equation (7.5.10) is equivalent to

$$K(x - y) = \pi \int_0^\infty q(t, x, y) \, dt + s(x, y) \tag{7.5.11}$$

for almost all y. To show that (7.5.11) holds for all y, write for $\epsilon > 0$, $\int_0^\infty q(t, x, y) \, dt$ in the form $\int_\epsilon^\infty q(t - \epsilon, x, y) \, dt$ and use (7.5.5):

$$\int K(x - z) p(\epsilon, z, y) \, dz = \pi \int_\epsilon^\infty q(t, x, y) \, dt \tag{7.5.12}$$

$$+ \int_0^\epsilon \mathbb{E}_x \left[p(s, X_T, y) \right] ds + \int s(x, y) p(\epsilon, z, y) \, dz$$

for all y. To see what happens as t tends to zero, we need the following:

Proposition 7.5.2. *For a constant A,*

$$\int_0^1 p(s, z, y) \, ds \leq A K(z - y) \quad \text{if } |z - y| \leq 1/2. \tag{7.5.13}$$

If m is a finite measure on $\mathbb{R}^2$ for which

$$u(y) = \int K(y-z) \, m(dz) \tag{7.5.14}$$

is superharmonic, then

$$\lim_{\epsilon \to 0} \int u(z) \, p(\epsilon, z, y) \, dz = u(y). \tag{7.5.15}$$

We postpone the proof and go ahead with our discussion. Since $p(s, z, y)$ is bounded on the set $\{s > 0, |y - z| \geq 1/2\}$, we see from (7.5.13) that

$$\int_0^1 \mathbb{E}_x \left[p(s, X_T, y) \right] ds < \infty$$

provided $\mathbb{E}_x \left[\|K(X_T - y)\| \right] < \infty$, that is, provided $\mathbb{E}_x \left[K(X_T - y) \right] < \infty$ because by Proposition 7.2.1, $\mathbb{E}_x \left[K(X_T - \cdot) \right]$ is superharmonic and hence by Exercise 6.1.8, $\mathbb{E}_x \left[K^-(X_T - y) \right] < \infty$ for all y. Thus, if $s(x, y) < \infty$, the middle term on the right side of (7.5.12) tends to zero as ϵ tends to zero. Now use Proposition 7.5.2 together with Proposition 7.2.1 to show that (7.5.11) is valid for all y such that $s(x, y) < \infty$. If $\infty = s(x, y) \; (\leq K(x-y))$, (7.5.11) is trivial. Since $s(x, y) = \infty$ only if $x = y$, we see from (7.5.11) that V defined by

$$V(x, y) = \int_0^\infty q(t, x, y) \, dt, \quad x, y \in \mathbb{R}^2, \tag{7.5.16}$$

is symmetric in (x, y). As is clear from (7.5.11), for $x, y \in W$, V is its Green function.

Proof of Proposition 7.5.2. To prove (7.5.13), we must show ($|z-y|^2 = r$)

$$A \log \frac{1}{r} - \int_0^1 \frac{1}{\pi t} \exp\left(-\frac{r}{2t}\right) dt \geq 0, \quad r \leq \frac{1}{2}$$

or calling $\frac{1}{r} = s$ and changing variables, it is needed to show

$$A \log s - \int_0^s \frac{1}{\pi t} \exp\left(-\frac{1}{2t}\right) dt \geq 0, \quad s \geq 2. \tag{7.5.17}$$

Differentiation of the above expression relative to s leads to

$$A/s - (1/\pi s) \exp(-1/2s)$$

which is certainly non-negative provided $A - 1/\pi \geq 0$. So the left side of (7.5.17) is increasing in s. If we choose $A \geq 1$ so that the left side of (7.5.17) is non-negative at $s = 2$, it will remain so for all $s \geq 2$. For such A, then (7.5.13) is established.

To prove (7.5.15), note first that

$$\int_{B(y,1)} u(z) p(\epsilon, z, y) \, dz \quad \text{tends to } u(y) \tag{7.5.18}$$

as ϵ tends to zero where $B(y, 1)$ is, as usual, the ball of radius 1 and center y. This is so because u being locally integrable, the existence of the integral (7.5.18) is clear. Using superharmonicity of u and integrating relative to polar coordinates show that the integral in (7.5.18) cannot exceed $u(y)$ because the integral of $p(\epsilon, z, y)$ over $B(y, 1)$ is at most one. On the other hand, lower semi-continuity of u and the fact that the integral of $p(\epsilon, z, y)$ over $B(y, 1)$ tend to 1 as ϵ tends to zero show that (7.5.18) is valid. Now

$$\int_{|y-z| \geq 1} p(\epsilon, z, y) |u(z)| \, dz \leq \int m(dx) \int_{|y-z| \geq 1} |K(x-z)| p(\epsilon, z, y) \, dz. \tag{7.5.19}$$

If $|y - z| \geq 1$, then $p(\epsilon, z, y) \leq 2p(1, z, y)$ for all $\epsilon > 0$. So the inner integral on the right side of (7.5.19) is dominated by

$$2 \int |K(a-z)| p(1, z, 0) \, dz = I_1 + I_2, \text{ say}, \quad a = x - y, \tag{7.5.20}$$

where I_1 is the integral over $B(a, 1)$ and I_2 over its complement. Clearly,

$$I_1 \leq \int_{|b| \leq 1} |K(b)| \, db = O(1). \tag{7.5.21}$$

The estimate (recall $a = x - y$) $\log |a - z| \leq \log(1 + |a|) + \log(1 + |z|)$ gives

$$I_2 \leq O(1) + \log(1 + |x - y|). \tag{7.5.22}$$

Now we can show that the right side of (7.5.19) tends to zero as ϵ tends to zero. Indeed, the inner integral on the right side of (7.5.19) tends to zero for every x (y fixed) because the integrand does so for all $z \neq x$ and is bounded by the (obviously integrable) integrand on the left side of (7.5.20). Further, the inner integral on the right side (7.5.19) is, as a function of x, bounded by $I_1 + I_2$ which is m-integrable as seen by the estimates (7.5.21) and (7.5.22). Recall that (7.3.2) must be valid if u is to be superharmonic.

Symmetry of q. In Chapter 8, we use that q is symmetric in x, y. We now prove this fact. Put for $\alpha \geq 0$

$$R_\alpha(x, y) = \int_0^\infty e^{-\alpha t} q(t, x, y) \, dy \tag{7.5.23}$$

so that R_0 is symmetric. By the semigroup property of q,

$$R_0 = R_\alpha + \alpha R_\alpha R_0 = R_\alpha + \alpha R_0 R_\alpha, \qquad (7.5.24)$$

where $R_\alpha R_0(x,y) = \int R_\alpha(x,z) R_0(z,y) \, dz$ and similar meaning is given to $R_0 R_\alpha$. The resolvent Equation (7.5.24) reveals a nice unicity property of R_0: If f and g are non-negative measurable and $\alpha > 0$,

$$f + \alpha R_0 f = g + \alpha R_0 g \qquad (7.5.25)$$

implies $R_0 f = R_0 g$ and hence $f = g$ at every point at which $R_0 f < \infty$. To see this, just operate by R_α both sides of (7.5.25) and use (7.5.24).

Now it is easy to prove the symmetry of q. Put $\hat{q}(t,x,y) = q(t,y,x)$ and define $\hat{R}_\alpha$ as in (7.5.23) replacing q by $\hat{q}$. The validity of (7.5.7) for all x, y shows that $\hat{q}$ also has the semi-group property. Using (7.5.24) and symmetry of R_0, i.e., $R_0 = \hat{R}_0$,

$$R_\alpha + \alpha R_0 R_\alpha = R_0 = \hat{R}_0 = \hat{R}_\alpha + \alpha \hat{R}_0 \hat{R}_\alpha$$

which by the unicity shown above leads to

$$R_\alpha(x,y) = \hat{R}_\alpha(x,y) = R_\alpha(y,x) \qquad (7.5.26)$$

at every point at which $R_0(x,y) < \infty$, i.e., for every x, y with $x \neq y$. For $x = y$, (7.5.26) being trivial, an appeal to the uniqueness of Laplace transform and a glance at (7.5.8) gives us the symmetry of q:

$$q(t,x,y) = q(t,y,x), \quad t > 0, \ x,y \in \mathbb{R}^d. \qquad (7.5.27)$$

For more examples and methods, consult [3, 9].

References

[1] G. Birkhoff and G.-C. Rota. *Ordinary Differential Equations*, 2nd edn. Blaisdell Publishing Co. [Ginn and Co.], Waltham, 1969.

[2] R. Courant and D. Hilbert. *Methods of Mathematical Physics*, Vol. I. Interscience Publishers, Inc., New York, 1953.

[3] G. F. D. Duff and D. Naylor. *Differential Equations of Applied Mathematics*. John Wiley & Sons, Inc., New York, 1966.

[4] F. Gesztesy and R. Nichols. Weyl-Titchmarsh M-functions for φ-periodic Sturm-Liouville operators in terms of Green's functions. In *From Complex Analysis to Operator Theory: A Panorama*. Operator Theory: Advances and Applications, Vol. 291, pp. 573–608. Birkhäuser/Springer, Cham, 2023.

[5] J.-C. Jiang, H.-W. Kuo, and M.-H. Liang. Green's function for solving initial-boundary value problem of evolutionary partial differential equations. *Int. J. Math.*, 34(9): 43, 2023. Paper No. 2350049.

[6] M. Marcus. Estimates of Green and Martin kernels for Schrödinger operators with singular potential in Lipschitz domains. *Ann. Inst. H. Poincaré C Anal. Non Linéaire*, 36(5): 1183–1200, 2019.

[7] M. Mayer and C. B. Ndiaye. Asymptotics of the Poisson kernel and Green's functions of the fractional conformal Laplacian. *Discrete Contin. Dyn. Syst.*, 42(10): 5037–5062, 2022.

[8] H. P. McKean, Jr. *Stochastic Integrals*. Probability and Mathematical Statistics, No. 5. Academic Press, New York, 1969.

[9] P. M. Morse and H. Feshbach. *Methods of Theoretical Physics*, 2 Vols. McGraw-Hill Book Co., Inc., New York, 1953.

[10] M. Mourgoglou. Regularity theory and Green's function for elliptic equations with lower order terms in unbounded domains. *Calc. Var. Partial Differ. Eqn.*, 62(9): 2023. Paper No. 266.

[11] C. G. Pacheco. Green kernel for a random Schrödinger operator. *Commun. Contemp. Math.*, 18(5): 1550082, 10, 2016.

[12] A. Rasheed, K. A. Khan, J. Pečarić, and Ð. Pečarić. Generalizations of Levinson type inequalities via new Green functions with applications to information theory. *J. Inequal. Appl.*, 19, 2023. Paper No. 124.

[13] C. Rejeb. Green function and Poisson kernel associated to root systems for annular regions. *Potential Anal.*, 55(2): 251–275, 2021.

[14] G. Riva, P. Romaniello, and J. A. Berger. Multichannel Dyson equation: Coupling many-body Green's functions. *Phys. Rev. Lett.*, 131(21): 2023. Paper No. 216401.

[15] A. Zygmund. *Trigonometric Series*, 2nd edn., Vols. I, II. Cambridge University Press, London, 1968. Reprinted with corrections and some additions.

[16] K. Knopp. *Problem Book*, Vol. 11. Dover Publications, Inc., New York, 1952.

Chapter 8

Potential Theory, Capacity, Boundaries, Dirichlet Spaces, and Applications

Historically, potential theory refers to the fundamental forces of nature, here referring to time as it arises in the theory of gravity and the electrostatic force. A successful model was first developed with the use of functions called the gravitational potential and electrostatic potential, both of which satisfy Poisson's equation — or in the vacuum, Laplace's equation. With the emergence of stochastic analysis, this has now turned into an exciting topic in its own right. Indeed, modern potential theory is now intimately connected with probability, and in particular, with the theory of Markov chains.

The first half of this chapter covers the two key notions of capacity and balayage. We add a comment on the latter: In potential theory, the term balayage (from French: balayage "scanning, sweeping", or more precisely "sweeping to the boundary") covers the following important step: It is a method devised initially by Henri Poincaré for the reconstruction of a given harmonic function in a domain from its values on the boundary of the domain. But it also serves the purpose of making precise a variety of potential theoretic boundaries. More precisely, it serves as a tool for computing harmonic measures and related kernels. In addition to their fundamental role in potential theory, balayage also plays a significant part in the fields of Coulomb gas and random matrix theory. For further reading on these applications, the reader may refer to the studies listed in references [1, 3, 14, 49, 51].

The original material in Chapter 7 dealing with potential theory is still current, and we have added a discussion of new directions. In addition, for the benefit of readers, we offer the following citations covering new directions [8–10, 23, 24, 26, 28–33, 41, 45, 46, 50, 52].

The new directions include randomized stopping and singular control; convergence of Dirichlet functions along random walk and diffusion paths; stochastic integration for volatility modulated Lévy driven Volterra processes; logarithmic potential theory and large deviation; analysis on infinite graphs; harmonic maps between Riemannian polyhedral; and capacity, convexity, and isoperimetric inequalities.

For the benefit of beginner readers, we offer the following supplementary introductory text [35].

While the number and variety of traditional applications of potential theory are well documented and substantial, it is of interest to also call attention here to more familiar settings. Case in point: Referring here to a recent textbook treatment, covering core graduate level complex analysis. Conway [17] has its Chapter 21 entirely devoted to potential theory in the (complex) plane, and, in special cases, it covers many of the key notions from potential theory, thus serving to illustrate uses of the general tools presented here, in the following.

Introduction. In this chapter, W will be a Greenian open set in $\mathbb{R}^d$, with Green function G.

In Section 8.1, we state and prove some of the main principles of potential theory. Section 8.2 is devoted to the celebrated capacity theorem some of whose applications are found in Section 8.3. Section 8.4 deals with the balayage procedure. In Section 8.5, we give the rudiments of Dirichlet spaces. Constraints of resources like time and space have forced us to drop subjects, such as additive functionals, Martin boundary, and fine topology.

8.1 Some Potential Theoretic Principles

Let s be excessive in a Green domain W, with Green function G:

$$h(x) = \lim_{D \uparrow W} \mathbb{E}_x\left[s(X_T)\right], \tag{8.1.1}$$

D relatively compact open in W,

$T = $ exit time from D,

is locally integrable and satisfies the mean value property and hence is harmonic in W. Also if u is harmonic in W and $u \leq s$, then clearly

$u \leq \mathbb{E}. [s(X_T)]$ for all relatively compact open D so that $u \leq h$:

h defined by (8.1.1) is the largest harmonic minorant of s. (8.1.2)

$s - h$ is excessive in W and its largest harmonic minorant must be zero. An excessive function whose largest harmonic minorant is zero is called a *potential*. Thus,

every excessive function can be written in a unique way as the sum of a harmonic function and a potential. (8.1.3)

To see uniqueness, suppose $s = h_1 + p_1 = h_2 + p_2$, where h_i are harmonic and p_i potentials. Then, $p_1 = p_2 + (h_2 - h_1) \geq h_2 - h_1$. Since p_1 is an potential $h_2 - h_1 \leq 0$ and by symmetry $h_1 - h_2 \leq 0$. Thus, $h_1 = h_2$ and $p_1 = p_2$.

There are no potentials on $\mathbb{R}^2$. Every excessive function on $\mathbb{R}^d$, $d \geq 3$, is the sum of a constant and a potential because non-negative harmonic functions on $\mathbb{R}^d$ are constants.

If s is excessive, then by Theorem 7.1.6,

$$s = \frac{1}{A_d} Gm + h, \qquad (8.1.4)$$

where m is the Riesz measure of s, $h \geq 0$ is harmonic, and

$$Gm(x) = \int G(x,y) \, m(dy). \qquad (8.1.5)$$

It is now clear that s is a potential if and only if

$$s = \frac{1}{A_d} Gm. \qquad (8.1.6)$$

Indeed, if s is a potential, (8.1.6) is a consequence of (8.1.4) since h has to be zero. Conversely, the Riesz measure of the excessive function $s = \frac{1}{A_d} Gm$ is m (as we saw before Theorem 7.1.6) and from (8.1.3), $s = p + h$ with h harmonic and p a potential. From what we said above, $p = \frac{1}{A_d} Gm$ and hence h must be zero and $s = p$.

We will be needing

$$G(x,y) = \mathbb{E}_x [G(X_H, y) : H < R], \quad y \in D, \ x \in W, \qquad (8.1.7)$$

where D is an open set, H = hitting time to D, and R = exit time from W.

To prove (8.1.7), start with the fact that $G(\cdot, y)$ is excessive harmonic except at y. So if y is fixed, the right side of (8.1.7) cannot become larger as $D \ni y$ becomes smaller. So assume that D is relatively compact in W.

For any relatively compact open A containing $\overline{D}$, $G(\cdot, y)$ is harmonic in the open set $A\backslash\overline{D}$ and continuous on its closure. Therefore, for all $x \in A\backslash\overline{D}$,

$$G(x,y) = \mathbb{E}_x\left[G(X_H, y) : H < S\right] + \mathbb{E}_x\left[G(X_S, y) : S < H\right],$$

where S = exit time from A. Let A increase to W. Using (8.1.2), because $G(\cdot, y)$ is a potential, we obtain (8.1.7) for all $x \notin D$. For $x \in D$, (8.1.7) is trivial.

Integrating both sides of (8.1.7) with respect to a measure m living on D:

$$\begin{cases} \text{If } Gm = p \text{ for a measure } m, D \text{ open and } m(W \backslash D) = 0, \text{ then} \\ \qquad p = \mathbb{E}.\left[p(X_H) : H < R\right] \qquad\qquad (8.1.8) \\ H = \text{hitting time to } D. \end{cases}$$

In this chapter, we are concerned mostly with potentials. We start with some of the main principles of potential theory. Note that the statements of these principles (except the domination principle of Maria–Frostman) are designed to make sense if the Green function G is replaced by any "Kernel", i.e., a positive lower semicontinuous function on $W \times W$.

The main principle of potential theory. In the following, W is a Greenian domain with Green function G, and for a measure m, Gm is the potential of m:

$$Gm(x) = \int G(x, y)\, m(dy).$$

The continuity principle. If Gm is continuous on support m, then Gm is continuous everywhere.

The minimum principle. The minimum of two potentials Gm_1 and Gm_2 is a potential Gm.

The uniqueness principle. If $Gm = Gn \not\equiv \infty$, then $m = n$.

The first or Frostman maximum principle. For all $x \in W$,

$$Gm(x) \leq \sup Gm(y),$$

where the supremum is over the support of m.

Domination or second maximum principle. If Gm is m-almost everywhere finite and

$$Gm \leq Gn, \quad m\text{-almost everywhere,}$$

then $Gm \leq Gn$ everywhere.

Potential Theory, Capacity, Boundaries, Dirichlet Spaces, and Applications   187

The last two principles are formulations, for measures, the "principle of positive maximum" of Section 4.3, and the last as the reader might have guessed says that "$\Delta u \leq 0$ at a maximum of u". The last two principles are implied by the *domination principle of Maria–Frostman.* If u is excessive, $Gm < \infty$ m-almost everywhere, and $u \geq Gm$ m-almost everywhere, then $u \geq Gm$ everywhere.

Proof. Let $p = Gm$. The hypothesis imply that we can find compacts F_n such that $u \geq p$ on F_n and the measures $m_n = m|F_n$ increase to m. Therefore these is no loss of generality in assuming that m has compact support F, that $u \geq p$ on F, and that $p < \infty$ m-almost everywhere.

We claim
$$p(\cdot) = \mathbb{E}.\,[p(X_H) : H < R], \tag{8.1.9}$$
$$H = \text{hitting time to } F$$
$$R = \text{exit time from } W.$$

Indeed, if D is relatively compact open and contains F and $D_1 = D\setminus F$, by Lemma 6.5.1,
$$p(\cdot) = \mathbb{E}.\,[p(X_T)], \quad T = \text{exit time from } D.$$
Since $T = H$ or $T = S$, where $S = $ exit time from D,
$$p(\cdot) = \mathbb{E}.\,[p(X_T) : H < S] + \mathbb{E}.\,[p(X_S) : S < H]$$
$$\leq \mathbb{E}.\,[p(X_T) : H < R] + \mathbb{E}.\,[p(X_S)].$$
Since p is a potential, the last term in the above inequality tends to zero as D increases to W by (8.1.2) and we obtain (8.1.9).

Since $X_H \in F$, if $H < \infty$, $u \geq p$ on F, and $u \geq \mathbb{E}.\,[u(X_H) : H < R]$, the proof now follows using (8.1.9). □

For the following principle, we need some preparation. We say that a Borel set A is quasi null if the following holds:

$m(A) = 0$ for every measure m such that Gm is uniformly bounded.

(8.1.10)

Clearly, a Borel set is quasi null if and only if every compact subset is and a countable union of quasi null sets is quasi null. We see in Section 8.3 that quasi null and polar are the same. In view of the maximum principle, a set A is quasi null if and only if $m(A) = 0$ for every measure m such that $Gm < \infty$ m-almost everywhere.

Let F be a compact set and H = hitting time to F. A point $x \in F$ is called *regular* (*irregular*) *for* F if
$$\mathbb{P}_x[H=0] = 1 \quad (=0).$$

Proposition 8.1.1. *The set of irregular points in a compact set F is quasi null.*

Proof. Just as in Proposition 5.2.3, $\mathbb{P}.[H \geq t]$ is seen to be upper semicontinuous. The set of irregular points in F is the union
$$\bigcup_n \left\{ x \in F : \mathbb{P}.\left[H \geq \frac{1}{n}\right] \geq \frac{1}{n} \right\} \tag{8.1.11}$$
and each set of the union is compact and consists entirely of irregular points.

Now suppose A is compact and not quasi null.

There then exists a measure m whose potential $p = Gm$ is m-almost everywhere finite and $m(A) > 0$. By restricting m to A if necessary, we may as well suppose that m lives on A. From (8.1.9) (H denotes hitting time to A and R = exit time from W),
$$\int G(x,y)\, m(dy) = p(x) = \mathbb{E}_x[p(X_H) : H < R]$$
$$= \int \mathbb{E}_x[G(X_H, y) : H < R]\, m(dy)$$
showing that for m-almost all y in A,
$$G(x,y) - \mathbb{E}_x[G(X_H, y) : H > R] = 0. \tag{8.1.12}$$

Just as in the "barrier" argument of Property 2, Section 7.1, one concludes from (8.1.12) that m-almost $y \in A$ are regular for A.

Finally, since each set of the union (8.1.11) consists entirely of irregular points, each of these sets and hence the union are quasi null. □

As an application, let us take up the matter of representing measures for $D(X)$, the set of continuous functions on the compact set X which are harmonic in the interior of X. Suppose a measure m on ∂X represents a point $x_0 \in X$, i.e., $\int f(y)\, m(dy) = f(x_0)$ for all $f \in D(X)$. We show that if m has no mass x_0, then m must be the harmonic measure at x_0. Consider $Km(\cdot) = \int K(\cdot - y)\, m(dy)$. If $x \notin X$, $K(x - \cdot)$ is harmonic in a neighborhood of x. Hence, $Km(x) = K(x - x_0)$. So as $x \notin X$ tends to a point $z \in \partial X$, by Fatou, $Km(z) \leq K(z - x_0)$. If m has no mass at x_0, Km is thus finite m-almost everywhere and hence cannot charge the set of irregular points in ∂X, i.e., it must live on the regular points in ∂X. That such a measure is the harmonic measure has already been shown in Section 6.5. An example of a harmonic measure with infinite energy (this notion is introduced later in this section) is given in Example 5.2.9.

Proposition 8.1.2. *Let Gn be n-almost everywhere finite. If $Gm \leq Gn$, then Gm is m-almost everywhere finite.*

Proof. Put $g = Gn$. If the m-measure of the set $(q = \infty)$ is not zero, by restriction if necessary let us assume that m lives on a compact subset F of $(q = \infty)$. For any open set containing F, if $p = Gm$, from (8.1.8),

$$p(x) = \mathbb{E}.\,[p(X_H) : H < R] \leq \mathbb{E}.\,[q(X_H) : H < R]. \tag{8.1.13}$$

That the last quantity in (8.1.13) tends to zero as D decreases to F for any x such that $g(x) < \infty$ is shown as follows:

If $\overline{D}$ is contained in the open set $(q > N)$,

$$N\mathbb{P}_x\,[H < R] \leq \mathbb{E}_x\,[q(X_H) : H < R] \leq q(x). \tag{8.1.14}$$

The best quantity in (8.1.13) is equal to

$$\int \mathbb{E}_x\,[G(X_H, y) : H < R]\,n(dy)$$

which decreases to zero as D decreases to F because we have the following: The integral is bounded above by the n-integrable function $G(x, \cdot)$, for every y not in F, it decreases to zero by (8.1.14), and n does not charge F. Thus, p must be zero, which can only be true if $m(F) = 0$. That proves the proposition. □

The balayage principle. Let $F \subset W$ be compact. For any measure m such that $p = Gm$ is m-almost everywhere finite, there exists a unique measure n, with support in F such that $Gm \geq Gn$ in W, while $Gm = Gn$ quasi everywhere on F (i.e., except for a quasi null subset of F). If $H =$ hitting time to F, $q(\cdot) = \mathbb{E}.\,[p(X_H) : H < R]$ is excessive in W, harmonic off F, is less or equal to p, and equals p at every point which is regular for F. If q_1 is any other potential with the properties $q_1 \leq p$ and $q_1 = p$ quasi everywhere on F, then $q_1 = q$ quasi everywhere on F because by Proposition 8.1.1 the set of irregular points in F is quasi null. The domination principle together with Proposition 8.1.2 implies that $q \equiv q_1$. Clearly, the Riesz measure of q is the required measure. n is called the *balayage* of m onto F.

The equilibrium principle. Any compact set E containing at least one regular point contains the support of a unique measure m whose potential $p = Gm \leq 1$ everywhere in W and $p = 1$ quasi everywhere on F. m is called *the equilibrium distribution on F and p its equilibrium potential.*

If $R =$ exit time from W and $H =$ hitting time to F,

$$p(\cdot) = \mathbb{P}.\,[H < R]$$

is easily seen to be a potential. [If $d = 2$, use the fact that $\mathbb{P}.\,[R < \infty] \equiv 1$. If $d \geq 3$, use the transience of the Brownian motion, c.f. Exercise 5.1.3.] p is then the equilibrium distribution. The uniqueness follows as before from the domination principle.

In physics, it is known that charges in a charged body redistribute themselves into an equilibrium state, and in this state (since there is no movement of charges), all points of the body are at the same potential. To find this distribution of charges is the equilibrium problem. The equilibrium principle asserts the existence and uniqueness of the equilibrium distribution.

The probabilistic meaning of the equilibrium distribution is explained by what is known as Chung's formula which we now describe. For a compact set F, the last exit time γ is defined as follows:

$$\gamma = \sup\{t : 0 < t < R,\ X_t \in F\}$$
$$= 0 \quad \text{if there is no such } t.$$

γ is not a stopping time but γ is measurable because $\{\gamma \leq t\} = \{H(\theta_t) \geq R\}$, where $H =$ hitting time to F and the shift operators θ are defined in Section 3.1. The last exit time satisfies $t + \gamma(\theta_t) = \gamma$ on the set $\gamma \geq t$ and the set $\{0 < \gamma\}$ is the same as the set $\{H < R\}$. For each subset $A \subset F$, the function $L(\cdot, A) = \mathbb{P}.\,(X_\gamma \in A,\ 0 < \gamma)$ is excessive (use Markov property) and (being less or equal to the equilibrium potential of F) is in fact a potential. Also, if A is compact, the hitting time to A is less or equal to γ on the set $\{X_\gamma \in A\}$. This fact and strong Markov property show that $L(\cdot, A)$ is harmonic off A, i.e., the Riesz measure of $L(\cdot, A)$ is concentrated on A if A compact. The sum of the potentials $L(\cdot, A)$ and $L(\cdot, F\backslash A)$ is the hitting potential p of F. Therefore, the sum of their Riesz measures is the Riesz measure of $p =$ the equilibrium measure m of F. In particular, $m(A, dy)$ ($=$ Riesz measure of $L(\cdot, A)$) $\leq m(dy)$ and, because the former is concentrated on A, we indeed have $m(A, dy) \leq 1_A(y) m(dy)$. We have thus shown that for every compact subset A of F,

$$L(\cdot, A) \leq \int G(\cdot, y) 1_A(dy) m(dy).$$

But the above being an identity when $A = F$, it must be an identity for all compact (and hence Borel) A contained in F. The proof of the identity

$$L(x, dy) = G(x, y) m(dy),$$

known as Chung's formula, is thus complete.

The energy principle. Physically speaking, $Gm(x)$ is the potential at x due to a distribution m of charges, i.e., $Gm(x)$ is the work needed to bring a point charge from infinity to x. Thus, $\int Gm(x) m(dx)$ is the total work needed to assemble this distribution of charges; in other words, it is the energy contained in this distribution of charges.

If m and n are (positive) measures, their *mutual energy* (m, n) is defined by

$$(m, n) = \int G(x, y) m(dx) n(dy) = (Gm, n) = (Gn, m), \quad (8.1.15)$$

where the symbol (Gm, n) has the obvious meaning that the potential Gm is integrated relative to n. We write $\|m\|^2$ for (m, m); this quantity is called the energy of m. The energy principle then states the following:

If m and n have finite energy, $(m, m) + (n, n) \geq 2(m, n)$ (8.1.16)

with equality only if $m = n$.

Only strict positivity, symmetry and lower semi-continuity of G are needed to establish the following useful proposition.

Proposition 8.1.3 (Gauss–Frostman). *Let F be a compact subset of W and f continuous on F. There is a measure m which minimizes*

$$Q(n) = (Gn, n) - 2(f, n) \quad (8.1.17)$$

among all positive measures n on F. $Gm = f$ on the support of m and the set of $x \in F$, where $Gm(x) < f$ has n-measure zero for every n on F of finite energy.

Proof. Since G is strictly positive and lower semicontinuous, its minimum of $F \times F$ is strictly positive. Therefore, (Gn, n) is of the order of magnitude $n(F)^2$ if $n(F)$ is large, whereas (f, n) is bounded by $(\sup |f|) n(F)$. Thus, $\inf Q(n) = \inf \{Q(n), n(F) < \Lambda\}$, where Λ is some large number. The set $\{n : n(F) \leq \Lambda\}$ is compact in the weak topology on measures on F ($C(F)$ is a Banach space and its dual is the set of all signed measures of finite total variation). G is lower semicontinuous and f is continuous so Q is a lower semicontinuous function of n and such a function attains its minimum on a compact set. Thus, there is a measure m minimizing Q.

Let m' be the restriction of m to the set $\{x \in F : Gm > f\}$. For $0 < \epsilon < 1$, $m - \epsilon m'$ is a (non-negative) measure on F. By choice of m, $Q(m - \epsilon m') - Q(m) \geq 0$ which is easily reduced to $\epsilon^2 (Gm', m') + 2\epsilon (f, m') - 2\epsilon (Gm, m') \geq 0$. Divide by ϵ and let ϵ tend to zero to get $(f, m') - (Gm, m') \geq 0$, which cannot be true unless $m' = 0$, because m' lives on the set $\{Gm > f\}$. Thus, m-almost everywhere, and hence by lower semi-continuity, on the support of m, $Gm \leq f$. In particular, m has finite energy.

Let now the measure n on F have finite energy and n' its restriction to the set $\{Gm < f\}$. n' has finite energy. The inequality $Q(m + \epsilon n') \geq Q(m)$ for all $\epsilon > 0$ leads as before to $(Gm - f, n') \geq 0$ which is false unless $n' = 0$. □

Proof of the Energy Principle. It is enough to prove (8.1.16) under the assumption that m, n live on a compact set F. Let $f \leq Gm$ be continuous on F. By Proposition 8.1.3, there is a measure μ on F such that $G\mu = f$ on support μ and

$$(Gn, n) - 2(f, n) \geq (G\mu, \mu) - 2(f, \mu) - - (G\mu, \mu) \tag{8.1.18}$$

because $G\mu = f$ on support μ. $G\mu = f \leq Gm$ on support μ implies by the domination principle $G\mu \leq Gm$ everywhere. Thus, $(G\mu, \mu) \leq (Gm, \mu) = (G\mu, m) \leq (Gm, m)$, which by (8.1.18) gives $(Gn, n) + (Gm, m) \geq 2(f, n)$. The validity of this for all continuous $f \leq Gm$ is just (8.1.16).

Thus, (8.1.16) is established except for showing that in case of equality m must equal n. For this note that a measure having C^1-density with compact support in W has finite energy. Let b be a measure with compact support with smooth density f. Then $m + b$ has finite energy ($(m, b) < \infty$ because Gm is locally integrable). Applying (8.1.16) to $m + b$ and n, and assuming that equality obtains in (8.1.16) for m and n, we get

$$(b, b) + 2(m, b) - 2(n, b) \geq 0.$$

Replacing b by ϵb,

$$\epsilon(b,b) + 2(m,b) - 2(n,b) \geq 0,$$

and letting $\epsilon \to 0$, we get

$$\int Gm(x) f(x) \, dx \geq \int Gn(x) f(x) \, dx$$

for all non-negative smooth functions f with compact supports. This implies $Gm \geq Gn$ almost everywhere and hence everywhere. By symmetry, $Gm = Gn$, i.e., $m = n$. Thus, (8.1.16) is completely established.

The above proof of the energy principle uses the domination principle which is very hard to check. Let us now indicate briefly another proof, which uses the result of Section 7.5 but not the domination principle.

A Second Proof of the Energy Principle. If q is the relative transition density (notation of Section 7.5), $\int q(t,\cdot,\cdot)$ is symmetric and has the semi-group property. Put for positive measures m and n, $Q(m,t,x) = \int q(t,x,y) m(dy)$ and $I(t,m,n) = \int q(t,x,y) m(dx) n(dy)$. Using the above properties of q, it is easily checked that

$$I(t,m,n) = \int Q(m,t/2,x) Q(n,t/2,x) \, dx$$

which in turn implies the following: $I(t,m,m) + I(t,n,n) \geq 2I(t,m,n)$, i.e., $q(t)$ is a definite kernel for each t by which we mean $I(t,\mu,\mu) \geq 0$ for any signed measure μ. Since $G = \int_0^\infty q(t) \, dt$, (8.1.16) is immediate.

The second method of proof gives us many kernels satisfying the energy principle. For example, $\int_0^\infty q(t,x,y) L(dt)$ satisfies the energy principle for almost any measure L. It would thus appear that this principle by itself is too general. However, together with the minimum principle — which is hard to check — it implies all the other principles. Take, for example, the domination principle. Suppose m and n have compact support and $Gm \leq Gn$ on support m. $q = Gm \wedge Gn$ is of the form Gl for some measure l. $(G(m-l), m-l) = (Gm,m) - (Gl,m) - (Gm,l) + (Gl,l) \leq 0$ because on support m, $Gm = Gl$ and $Gm > Gl$ everywhere. By the energy principle, $m = l$ and domination follows.

If in (8.1.16) we take tn instead of n and let $\|n\| t = \|m\|$, we get the *energy inequality*

$$(m,n) \leq \|m\| \|n\| \tag{8.1.19}$$

of which an immediate consequence is

$$\|m+n\| \le \|m\| + \|n\|. \tag{8.1.20}$$

Let ξ_+ denote the set of all measures of finite energy. For $m_1, m_2 \in \xi_+$, $m = m_1 - m_2$ can be regarded as a signed Radon measure, i.e., as a linear functional on the continuous functions of compact supports in W. $Gm = Gm_1 - Gm_2$ makes sense as an n-integrable function for every $n \in \xi_+$. Now we let ξ denote the set of all signed Radon measures $m = m_1 - m_2$ with $m_1, m_2 \in \xi_+$. Equations (8.1.19) and (8.1.20) make ξ a pre-Hilbert space with inner product $(m, n) = \int Gm\,dn$. *ξ is not complete.* The completion of ξ can be identified with a space of distributions; see Section 8.5. However, a result of Cartan states that ξ_+ is complete. We now prove this and indicate two consequences.

Lemma 8.1.4. *Let $\{s_i\}$ be a sequence of excessive functions and $s = \inf_i s_i$. Let $\hat{s}$ denote the lower semicontinuous regularization of s:*

$$\hat{s}(x) = \liminf_{y \to x} s(y).$$

Then $\hat{s}$ is excessive and the set $\{\hat{s} < s\}$ is quasi null.

Proof. Let $a > 0$. We show that every compact subset F of the set $\{s > \hat{s} + 2a\}$ consists entirely of irregular points. By Proposition 8.1.1, F and hence the set itself are quasi null.

Let $z \in F$. $\hat{s}$ being lower semicontinuous, there is a neighborhood V of z such that for all $x \in \overline{V}$,

$$\hat{s}(x) > \hat{s}(z) - a, \quad x \in \overline{V}. \tag{8.1.21}$$

Then, for all $x \in \overline{V} \cap F$ and all i,

$$s_i(x) \ge \hat{s}(x) + 2a > \hat{s}(z) + a = A \text{ say}, \quad x \in \overline{V} \cap F. \tag{8.1.22}$$

If $H = $ hitting time to $\overline{V} \cap F$, from (8.1.22) for all x,

$$s_i(x) \ge \mathbb{E}_x[s_i(X_H) : H < R] \ge A\mathbb{P}_x[H < R]. \tag{8.1.23}$$

The right side of (8.1.23) is independent of i and is lower semicontinuous. From the definition of $\hat{s}$,

$$\hat{s}(x) \ge A\mathbb{P}_x[H < R] = (\hat{s}(z) + a)\mathbb{P}_x[H < R], \quad x \in W.$$

The last inequality when $x = z$ excludes the possibility that z is regular for $\overline{V} \cap F$ and regularity being a local property, z cannot be regular for F.

To show that $\hat{s}$ is excessive, appeal to Proposition 6.1.2: If u is harmonic in a relatively compact open set D, continuous on $\overline{D}$, and $u \leq \hat{s}$ on ∂D, then $u \leq s_i$ in D for all i. Since it is continuous in the open set D, $u \leq \hat{s}$ as well. □

Lemma 8.1.5. *Let m_i be a sequence which is bounded in ξ_+ and $p_i = Gm_i$. Then there exists $m \in \xi_+$ whose potential p equals $\liminf p_i$ quasi everywhere.*

Proof. By Lemma 8.1.4, for each i, $\inf\{p_j : j \geq i\}$ is equal to a potential q_i quasi everywhere. $q_i \leq q_{i+1}$. $p = \lim q_i$ is then excessive and equals $\liminf p_i$ quasi everywhere. Let $x_0 \in W$, D be a relatively compact open set and T = exit time from D. Now $p \leq \liminf p_i$. So to show that p is a potential, we need only show that

$$\mathbb{E}_{x_0}\left[p_i\left(X_T\right)\right] \tag{8.1.24}$$

is uniformly small if D is large. If n is the harmonic measure at x_0 relative to D, i.e., $n(dy) = \mathbb{P}_{x_0}[X_T \in dy]$, the quantity in (8.1.24) is simply $\int p_i dn = (m_i, n)$ m_i being bounded, by (8.1.19), it is sufficient to show that the energy of n is small if D is large. The following shows precisely this:

$$\int G\left(x, y\right) n\left(dy\right) = \mathbb{E}_{x_0}\left[G\left(X_T, x\right)\right] \leq G\left(x_0, x\right)$$

$$(n, n) \leq \int G\left(x_0, y\right) n\left(dy\right) = \mathbb{E}_{x_0}\left[G\left(x_0, X_T\right)\right]$$

and the last quality tends to zero as D increases to W because $G(x_0, \cdot)$ is a potential.

Finally, it remains to show that the Riesz measure m of p is in ξ_+. By Fatou, $\int p\,dm \leq \liminf_i \int p_i dm$, and for each i, $\int p_i dm = \int p\,dm_i \leq \liminf_j \int p_j dm_i = \liminf_j (m_j, m_i)$, and (m_i, m_j) is uniformly bounded by (8.1.19) and the boundedness of m_i in ξ_+. □

Theorem 8.1.6. *ξ_+ is complete.*

Proof. Let $m_i \in \xi_+$ be a Cauchy sequence. Choosing a subsequence if necessary, assume $\sum \|m_i - m_{i+1}\| < \infty$. Put $p_i = Gm_i$. For $n \in \xi_+$,

$$\int |p_i - p_{i+1}|\,dn = \int (p_i - p_{i+1})\,dn_1 + \int (p_i - p_{i+1})\,dn_2 \tag{8.1.25}$$

$$= (m_i - m_{i+1}, n_1) + (m_{i+1} - m_i, n_2),$$

where n_1 and n_2 are the restrictions of n to the sets $\{p_i \geq p_{i+1}\}$ and $\{p_{i+1} \geq p_i\}$, respectively. $n_1, n_2 \in \xi_+$ and their energies are dominated by the energy of n. From (8.1.19) and (8.1.25),

$$\sum \int (p_i - p_{i+1}) \, dn \leq 2 \|n\| \sum \|m_i - m_{i+1}\| < \infty.$$

In particular, for any $n \in \xi_+$, $\lim p_i$ exists n-almost everywhere and, putting $s = \liminf p_i$,

$$\lim \int |s - p_i| \, dn = 0. \tag{8.1.26}$$

But by Lemma 8.1.5, there is $m \in \xi_+$ such that $s = Gm$ quasi everywhere. Equation 8.1.26 then implies (m_i, n) tends to (m, n), i.e., m_i tends weakly (in the Hilbert space sense) to m. Being a Cauchy sequence, it tends strongly to m. □

Before proceeding for applications, note the following. If a sequence $m_i \in \xi_+$ tends weakly to $m \in \xi_+$, then m_i tends vaguely to m. Indeed, if φ is C^2 with compact support,

$$-A_d \int \varphi \, dm_i = (\Delta \varphi, m_i) \to (\Delta \varphi, m) = -A_d \int \varphi \, dm \tag{8.1.27}$$

because $\Delta \varphi$ defines a signed measure of finite energy and $\int G(\cdot, y) \Delta \varphi(y) \, dy = -A_d \varphi$. Also the energies of m_i have to be uniformly bounded and so (G is strictly positive and lower semicontinuous), $m_i(F)$ is uniformly bounded for each compact set F. Vague convergence of m_i to m is thus clear from (8.1.27). *In particular, the set of measures of finite energy living on a closed set is a closed convex subset of ξ_+.*

With this preliminary, we can give a geometrical meaning to balayage: Let m be any element of ξ_+ and F a closed set. By elementary Hilbert space theory, there is a unique measure n on F — the projection of m on the closed convex set of measures of finite energy living on F — such that $\|m - n\|^2 \leq \|m - \mu\|^2$ for all $\mu \in \xi_+$ which live on F. Take $n + t\mu$ instead of μ and reduce to get

$$-2(m - n, t\mu) + t^2 \|\mu\|^2 \geq 0 \tag{8.1.28}$$

for all such μ. This can only be true if $(m - n, \mu) \leq 0$ and (taking $t = 1$, $\mu = n$ in (8.1.28)) $(m - n, n) = 0$. As in the proof of Proposition 8.1.3, this implies $Gm \leq Gn$ quasi everywhere on F. But then by $(m - n, n) = 0$, we must have $Gm = Gn$, n-almost everywhere. By the domination principle,

Potential Theory, Capacity, Boundaries, Dirichlet Spaces, and Applications   197

$Gm > Gn$ everywhere on W. Thus, $Gm > Gn$ everywhere on W and equality holds quasi everywhere on F, i.e., n is the balayage of m.

Similarly, the equilibrium distribution has the following interpretation. Let F be a compact set. The set of probability measures on F of finite energy is a closed convex subset of ξ_+. The projection of the zero measure on this set is simply the unique probability measure m on F of minimal energy. $\|m\|^{-2} m$ is the equilibrium distribution for F.

The existence parts of balayage and the equilibrium principles can be directly deduced from Proposition 8.1.3. See Exercises 8.1.6 and 8.1.7.

Remark. The reader might question the wisdom of proving a difficult theorem like the completeness of ξ_+ just to be able to interpret balayage as a projection. The point is, with this method, a sort of balayage can be worked out even when the domination principle fails. A case in point is this: The M. Riesz kernels $|x - y|^{-d+\alpha}$ satisfy the domination principle only when $0 < \alpha \leq 2$. But for all $0 < \alpha < d$, the corresponding ξ_+ is complete. Thus, for $2 < \alpha < d$, the Hilbert space method gives a solution to balayage problem, whereas other methods fail. We discuss these matters in a latter section.

Exercises

8.1.1. Show that the first maximum principle implies the continuity principle.

Hint: See the hint to Exercise 6.4.2.

8.1.2. Show that the second maximum principle implies the first maximum principle.

Hint: Suppose m has compact support. $Gm \leq 1$ on F. If D is an open set containing F, the hitting potential of D is equal to 1 on F.

Remark. That the continuity principle implies the Maria–Frostman maximum principle is shown in the text. Thus, these principles are equivalent.

8.1.3. Let q be a potential and $A = q^{-1}(\infty)$. Show that A is quasi null.

Hint: If F is a compact subset of A, m lives on F and $p = Gm$ is m-almost everywhere finite, then (8.1.9) holds so that the $\mathbb{P}_x$-probability of the event $\{H < R\}$ cannot be zero. On the other hand, if $q(x) < \infty$,

$$\mathbb{E}_x\left[q\left(X_H\right) : H < R\right] \leq q(x) < \infty \quad \text{and} \quad q = \infty \text{ on } F.$$

8.1.4. Let q and A be as in Exercise 8.1.3 above. Then for every measure n living on A, $Gn = \infty$, n-almost everywhere.

Hint: If F is compact $\subset A$, $n(F) > 0$, $Gn < \infty$ on F, and $m = n|F$, then $Gm < \infty$, m-almost everywhere. Now use Exercise 8.1.3.

8.1.5. Show that the set $m \in \xi$ such that m and Gm both are compactly supported is dense in ξ.

Hint: Let $m \in \xi_+$ have compact support, D relatively compact open, and T = exit time form D. If $p = Gm$, then $q = \mathbb{E}.\,[p(X_T)]$ is Gn for some n. Since $q \leq p$, $\|n\|^2 \leq \int p\,dn = \int q\,dm \leq \int p\,dm = \|m\|^2 < \infty$. As D increases to W, the m-integrable functions q decrease to 0 and are bounded by the m-integrable function p. So $\|n\|$ is small for large D. $m - n$ is the required signed measure. See also Lemma 8.5.1.

8.1.6. From Proposition 8.1.3 and the domination principle, derive the balayage principle.

Hint: Let F be compact and f continuous on F with $f \leq Gm$. There is a measure n on F such that $Gn = f$ on support n, $Gn \geq f$ quasi everywhere on F, and $Gn \leq Gm$ everywhere. Let f increase to Gm.

8.1.7. Using Proposition 8.1.3 and the first maximum principle, derive the equilibrium principle.

Hint: Take $f = 1$ in Proposition 8.1.3. For uniqueness, we need also the domination principle.

8.2 Capacity

Let F be compact $\subset W$ (our Green domain) and p its equilibrium potential:

$$p(\cdot) = \mathbb{P}.\,[H < R],$$

where H = hitting time to F and R = exit time from W. The Riesz measure of p is concentrated on ∂F because $p \equiv 1$ in the interior of F and is obviously harmonic off F. We call $m(F)$ the (Newtonian) capacity of F:

$$N(F) = m(F) = m(W) = m(\partial F).$$

Just as in measure theory we define the inner and outer capacities of our arbitrary subset A of W as follows:

The inner capacity $N_*(A) = \sup N(F)$,

where the supremum is over all compact subsets F of A.

The outer capacity $N^*(A) = \inf N_*(D)$,

where the infimum is over all open sets $D \supset A$. A set E is called *capacitable* if its inner and outer capacities coincide and $N(E)$ will denote this common value and is called its capacity.

Proposition 8.2.1. *Let a and b be Radon measures with potentials u and v. If $u \leq v$, then*

$$a(W) \leq b(W).$$

Proof. Let F be any compact subset of W and p its equilibrium potential and m its equilibrium measure. Then, if $F° = $ interior of F,

$$a(F°) \leq \int p\,da = \int u\,dm \leq \int v\,dm = \int p\,db \leq b(W)$$

because $p = 1$ on $F°$ and ≤ 1 everywhere. The proposition follows by letting F increase to W.

Observation. The domination principle of Section 8.1 and Proposition 8.2.1 above permit us to identify the capacity of a compact set as follows:

$$N(F) = \sup m(F),$$

where the sup is over all measures m on F whose potential $Gm \leq 1$ everywhere. In particular, a compact set has capacity zero if and only if it is quasi null. Now a Borel set is *quasi null* if and only if every compact subset is quasi null. And we obtain the following:

A Borel set is quasi null if and only if its inner capacity is zero.

That *a compact set is capacitable* is seen as follows: Let A be compact and D a relatively compact open set containing A. Suppose D_n are open, $\overline{D}_n \subset D$, and D_n decrease to A. Denote by p the relative hitting potential of D:

$$p(\cdot) = \mathbb{P}.\,[H < R], \quad H = \text{hitting time to } D,$$

and let p_n similarly denote the hitting potentials to D_n. The Riesz measure m of p is concentrated on ∂D and p_n decrease to the hitting potential q of A at all points of ∂D (in fact, $p_n(x)$ decreases to $q(x)$ for all x, except when $x \in \partial A$ is irregular). If m_n are the Riesz measures of p_n, because $p = 1$ on $\overline{D}_n$ and m_n live on ∂D_n,

$$m_n(W) = \int p\,dm_n = \int p_n\,dm \downarrow \int q\,dm = \int p\,da = a(W),$$

where a is the Riesz measure of q, i.e., the equilibrium distribution for A. On the other hand, it is clear from Proposition 8.2.1 that $m_n(W) \geq N_*(D_n)$. Therefore,

$$N^*(A) = N(A) = N_*(A), \quad A \text{ compact.} \tag{8.2.1}$$

Theorem 8.2.2. *The outer capacity N^* has the following properties:*

(i) $N^*(A) \leq N^*(B)$ if $A \subset B$.
(ii) $N^*(A_n) \downarrow N^*(A)$ if $A_n \downarrow A$ strongly, i.e., in such a way that every open set containing A contains some A_n.
(iii) $N^*(A \cup B) + N^*(A \cap B) \leq N^*(A) + N^*(B)$. This property is called strong subadditivity.
(iv) $N^*(A_n) \downarrow N^*(A)$ if $A_n \uparrow A$.
(v) Borel sets (even analytic sets) are capacitable.

Proof. (i) and (ii) are clear from the definition. Let us prove (iii). If A and B are compact and H_1, H_2 the hitting times to A and B respectively, $H_1 \wedge H_2$ is the hitting time to $A \cup B$ and $H_1 \vee H_2$ is less or equal to the hitting time H to $A \cap B$. If $R =$ exit time from W (recall $H \geq H_1 \vee H_2$), the trivial inequality

$$\mathbb{P}_x[H_1 \wedge H_2 < R] + \mathbb{P}_x[H < R] \leq \mathbb{P}_x[H_1 < R] + \mathbb{P}_x[H_2 < R]$$

leads by Proposition 8.2.1 to

$$N(A \cup B) + N(A \cap B) \leq N(A) + N(B). \tag{8.2.2}$$

If D_1 and D_2 are open,

$$N_*(D_1 \cup D_2) + N_*(D_1 \cap D_2) \leq N_*(D_1) + N_*(D_2) \tag{8.2.3}$$

follows from the definition of N_*, (8.2.1) and the elementary topological fact that if A and B are compact subsets of $D_1 \cup D_2$ and $D_1 \cap D_2$, respectively, then we can find compact sets $A_1 \subset D_1$ and $A_2 \subset D_2$ such that $A_1 \cup A_2 \supset A$ and $A_1 \cap A_2 \supset B$. [Write $D_1 = \cup A_n$, $D_2 = \cup B_n$ with A_n, B_n increasing open and relatively compact in D_1, D_2, respectively. For some n, $A_n \cup B_n$ will contain A and $A_n \cap B_n$ will contain B.] Now the proof of (iii) should create no problems.

To prove (iv), we may and do assume that $\sup_n N^*(A_n) < \infty$. Let $a > 0$. We find inductively an increasing sequence of open sets $D_n \supset A_n$ such that $N_*(D_n) < N^*(A_n) + a$. Let D_1 be open, $D_1 \supset A_1$, and

Potential Theory, Capacity, Boundaries, Dirichlet Spaces, and Applications   201

$N_*(D_1) < N^*(A_1) + Ca/2$. Suppose $D_1 \subset D_2 \subset \cdots \subset D_n$ have been found such that $D_n \supset A_n$ and $N_*(D_n) < N^*(A_n) + (1 - 2^{-m})a$. Let D^1_{n+1} be open with $N_*(D^1_{n+1}) < N^*(A_{n+1}) + a2^{-n-1}$. Put $D_{n+1} = D_n \cup D^1_{n+1}$, $D_{n+1} \supset A_{n+1}$. From (8.2.3),

$$N_*(D_{n+1}) + N_*(D^1_{n+1} \cap D_n)$$
$$\leq N_*(D_n) + N_*(D^1_{n+1})$$
$$\leq N^*(A_n) + N^*(A_{n+1}) + a(1 - 2^{-n} + 2^{-n-1}).$$

Since $D^1_{n+1} \cap D_n \supset A$, we get from the above inequality,

$$N_*(D_{n+1}) \leq N^*(A_{n+1}) + a(1 - 2^{-n-1}). \tag{8.2.4}$$

If $D = \cup D_n$, then D contains A. And $N_*(D) = \sup N_*(D_n)$ because any compact subset of D is contained in some D_n.

(v) is proved by using the famous capacity theorem which is as follows:
□

The capacity theorem. We need to introduce some terminology. Let us denote by $\mathcal{N}$ the topological product of countably many copies of the set of natural numbers $\{1, 2, 3, \ldots\}$ with the discrete topology. It is well known that $\mathcal{N}$ is homeomorphic to the set of irrationals in the interval $[0, 1]$ but we do not need this. See Kuralowski [1].

We need the following facts about $\mathcal{N}$:

(i) If $\underline{n} = (n_1, n_2, \ldots) \in \mathcal{N}$, then the sets

$$N(n_1, \ldots, n_k) = \{m \in \mathcal{N} : m_i = n_i,\ 1 \leq i \leq k\}$$

form a base of neighborhoods of $\underline{n}$ and each of these sets is open and closed.

(ii) A countable union of disjoint copies of $\mathcal{N}$ is homeomorphic to $\mathcal{N}$. Denoting the disjoint copies by $(k, \mathcal{N})$, the map, which sends $(k, \underline{n})$ into the element of $\mathcal{N}$ with first coordinate k and whose $(i+1)$-st coordinate is the ith coordinate of $\underline{n}$, is a homeomorphism of $\bigcup_k (k, \mathcal{N})$ onto $\mathcal{N}$.

(iii) Countable product of copies of $\mathcal{N}$ (being still a countable product of the natural numbers) is homeomorphic to $\mathcal{N}$.

A Hausdorff topological space is called a *Polish space* if there is a metric on it — consistent with the given topology — making it complete and separable. A countable product of Polish spaces is Polish. A Hausdorff topological space is called *analytic* if it is the continuous image of $\mathcal{N}$. The set of natural numbers being a Polish space, $\mathcal{N}$ and also any closed subset of $\mathcal{N}$ is Polish.

With these preliminaries, we have the following:

Lemma 8.2.3. *A Polish space is analytic. The class of analytic subsets of an analytic space is closed under countable unions and countable intersections and contains all Borel sets (i.e., sets in the σ-field generated by closed sets.)*

Proof. Let (Y, d) be a complete separable metric space. For each finite set $\{n_1, n_2, \ldots, n_k\}$ of integers, we find non-empty closed balls of diameter $\leq \frac{1}{k}$ in the following way: Let $F_1, F_2, \ldots, F_n, \ldots$ be a cover of Y with closed balls of diameter ≤ 1; in case of finite cover, we repeat the last one. If $F_{n_1 n_2 \cdots n_k}$ has been found, we cover $F_{n_1 n_2 \cdots n_k}$ by closed balls $F_{n_1, n_2 \cdots n_k j}$ of diameter $\leq \frac{1}{k+1}$ (repeating the last if necessary). $F_{n_1 \cdots n_k}$ thus found has the following properties:

$$\text{diameter } F_{n_1 \cdots n_k} \leq \frac{1}{k}, \quad \bigcup_{n_1 \cdots n_k} F_{n_1 \cdots n_k} = Y. \tag{8.2.5}$$

If $\underline{n} = (n_1, n_2, \ldots, n_k, \ldots) \in \mathcal{N}$, we define $f(\underline{n}) = \bigcap_k F_{n_1 \cdots n_k}$. This intersection is non-empty and contains exactly one point by completeness and Hausdorffness. That it is onto follows from (8.2.5). The set $N(n_1, \ldots, n_k) = \{q : \underline{q} = (q_1, \ldots, q_k, \ldots), q_i = n_i, i \leq k\}$ is open in $\mathcal{N}$ and $f^{-1}(F_{n_1 \cdots n_k})$ contains $N(n_1, \ldots, n_k)$. This proves continuity. Thus, Y is the continuous image of $\mathcal{N}$. Thus, a continuous image of a Polish space is analytic.

Let Y be analytic and A_i analytic subsets of Y. $\cup_i A_i$ will then be a continuous image of disjoint copies of $\mathcal{N}$ and union of countable disjoint copies of $\mathcal{N}$ is homeomorphic to $\mathcal{N}$.

To show that $\cap_i A_i$ is analytic, note first that a closed subset of an analytic space is analytic because it is a continuous images of a closed subset of $\mathcal{N}$ and the latter is polish. Now $\cap_i A_i$ can be identified with a closed subset of the product $\prod_i A_i$, namely the set of $x = (x_1, x_2, \ldots)$ with $x_i = x_j$ for all i and j. And the latter is analytic being the continuous image of the product $\prod_i \mathcal{N}$ (which is homeomorphic to $\mathcal{N}$).

We have also shown above that Borel sets are analytic because a family closed under countable unions and intersections and containing closed sets contains all Borel sets. $\square$

Some more information on analytic spaces is in exercises.

Let Y be an analytic space. By a Choquet capacity on Y, we mean a non-negative set function C defined on all subsets of Y such that

(1) $C(A) \leq C(B)$ if $A \subset B$,
(2) $C(A_n) \uparrow C(A)$ if $A_n \uparrow A$,
(3) $C(A) = \lim C(A_n)$ if A_n decreases strongly to A, i.e., $A = \cap A_n$ and each open set containing A contains some A_n.

We have the fundamental as follows:

Theorem 8.2.4 (Capacity Theorem). *Let Y be an analytic space and C a Choquet Capacity on Y. Then for any number $a < C(Y)$, there is a compact subset $A \subset Y$ such that $a \leq C(A)$.*

Proof. Let f be continuous from $\mathcal{N}$ onto Y. Let V_k denote the set of $\underline{n} \in \mathcal{N}$ whose first coordinate $\leq k$:

$$V_k = \{\underline{n} \in \mathcal{N} : \underline{n} = (n_1, \ldots),\ n_1 \leq k\}.$$

Since $V_k \uparrow \mathcal{N}$ as $k \uparrow \infty$, $f(V_k) \uparrow Y$. We can find k_1 such that $C(f(V_{k_1})) > a$. Let $V_{k_1 j}$ be the set of $\underline{n} \in V_{k_1}$ whose second coordinate is less or equal to j. $V_{k_1 j}$ increases to V_{k_1}. We can find k_2 such that $C(f(V_{k_1 k_2})) > a$. This procedure determines for each j a set $V_{k_1 \cdots k_j}$.

$$V_{k_1 \cdots k_j} = \{n : (n_1, \ldots, n_k, \ldots) \in \mathcal{N},\ n_1 \leq k_1, \ldots, n_j \leq k_j\}$$

such that $C(f(V_{k_1 \cdots k_j})) > a$. Put

$$V = \bigcap_j V_{k_1 k_j} = \{\underline{n} : n_i \leq k_i\}.$$

That V is compact and $V_{k_1 \cdots k_j}$ strongly decrease to v are easily shown. It follows that $A = f(V)$. We conclude $C(A) \geq a$. $\square$

Exercises

8.2.1. Show that a G_δ-subset of a Polish space is Polish.

Hint: As we saw in the proof of Lemma 8.2.3, intersection can be identified with a closed subset of a product. An open subset U of a Polish space is complete with the metric

$$d(x,y) + \left| \frac{1}{d(x,F)} - \frac{1}{d(y,F)} \right|,$$

where d is the metric in the ambient space and $F =$ the complement of U.

8.2.2. A topological space is a Lindelöf space if every open cover has a countable sub-cover. Show that every analytic space is a Lindelöf space.
 Hint: Continuous image of a Lindelöf space is Lindelöf.

8.2.3. Let A and B be analytic. Show that the Borel field in $A \times B$ is the product Borel field.
 Hint: $A \times B$ is analytic so is every open subset. An open set O in $A \times B$ is a union of open sets of the form $U \times V$ where U is open in A and V is open in B. There is a countable sub-cover by Exercise 8.2.2 and $U \times V$ belongs to the product Borel field.

Remark. Some result clearly holds for countable product of analytic spaces.

8.2.4. Let A and B be analytic and $f : A \to B$ Borel measurable. Show that graph f is a measurable subset of $A \times B$.
 Hint: The map $B \times A \xrightarrow{g} B \times B$ given by $g(b,a) = (b, f(a))$ is measurable and the inverse image of the diagonal in $B \times B$ (which is measurable relative to the product Borel field by Exercise 8.2.3) is simply the graph of f.

8.2.5. Show that images and inverse images under Borel maps of analytic sets are analytic.
 Hint: $f : X \to Y$ is Borel, $A \subset Y$ analytic, then
$$f^{-1}(A) = \Pi_x \left[(\text{graph } f) \cap (X \times A) \right],$$
and if $A \subset X$ is analytic, then
$$f(A) = \Pi_y \left[(\text{graph } f) \cap (A \times Y) \right],$$
where Π_x and Π_y are projections onto X and Y.

8.2.6. If D is an open relatively compact subset of W, the hitting function $u = \mathbb{P}.\,[H < R]$, $H = $ hitting time to D, is necessarily a potential. For D not relatively compact, give examples where u is harmonic, harmonic + potential, or just potential.

8.2.7. Let D be any open subset of W and u as in Exercise 8.2.6 above. If u is a potential, then $N(D)$ is the total mass of the Riesz measure of U. If u is not a potential, then $N(D) = \infty$.
 Hint: The energy of the equilibrium distribution of every compact subset of D is bounded by $N(D)$. So, if $N(D) < \infty$, by Lemma 8.1.5, u must be a potential. Of course, the same holds for analytic sets.

8.3 Applications

Our first application of the capacity theorem is to (v) of Theorem 8.2.2. (i), (ii), and (iv) of Theorem 8.2.2 say that N^* is a Choquet capacity. We have already shown that for a compact set $N^* = N$. Theorem 8.2.4 then guaranties that *all analytic sets are capacitable*.

Our next application is to showing that a Borel set is quasi null (definition in Section 8.1) if and only if it is polar. Recall the definition of a polar set given in Section 6.2: A set is called polar if it is contained in the set of poles, i.e., infinities of a superharmonic function. We need a small proposition in which we prove more (for future reference) than is immediately needed.

Proposition 8.3.1. *Let $F \subset W$ be polar and $x_0 \notin F$. Then there exists a potential p with finite Riesz measure such that $p \equiv \infty$ on F and $p(x_0) \leq 1$.*

Proof. Suppose first that F is relatively compact. F being polar there is a superharmonic function s which is identically infinity on F. Let n_1 be the Riesz measure of s. If D is relatively compact open neighborhood of $\overline{F}$, then $n_1(D) < \infty$. If $n = n_1|D$, the potential g of n and s have the same Riesz measure on D implying that $s = g + h$, with h harmonic in D. g is then infinite on F.

In the general case, $F = \cup_n F_n$ with F_n relatively compact. If g_n are determined as above, for suitable constants a_n, $g = \sum a_n g_n$ will be a potential with finite Riesz measure which is infinite on F.

Finally, let $x_0 \notin F$ and $B_n = B(x_0, 1/n)$ be balls with center x_0 and radius $1/n$. With T_n = exit time from B_n,

$$p_n(\cdot) = \mathbb{E}.\left[g(X_{T_n})\right]$$

are potentials, are finite at x_0 and equal g off B_n. In particular, $p_n \equiv \infty$ on $F \backslash B_n$. For suitable constants b_n, $p = \sum b_n p_n$ is a potential with finite Riesz measure, $p = \infty$ on F and $p(x_0) \leq 1$.

Now it is easy to show that a polar Borel set A is quasi null. If not, there would exist a measure m on A whose potential $p = Gm$ is bounded. Let g be a potential with finite Riesz measure n such that $g = \infty$ on A. We arrive at the contradiction

$$\infty = \int g dm = \int p dn < \infty.$$

In view of the observation made after Proposition 8.2.1, Theorem 8.2.4 asserts that a Borel set is quasi null if and only if it has capacity zero. Let us now show that *a set A of capacity zero is polar*.

We may assume A is relatively compact. Let $A \subset D_n$ be a sequence of relatively compact open sets, all contained in a fixed compact set, such that $N(D_n) \leq 2^{-n}$. Now $N(D_n)$ (c.f. Exercise 8.2.2) is simply the total mass of the Riesz measure m_n of the hitting potential p_n of D_n because the supremum of the hitting potentials of compact subsets of D_n is simply the hitting potential of D_n. $p = \sum p_n$ is a potential because $p = Gm$, where $m = \sum m_n$ is a finite measure and all the m_n live on the compact set in which all the D_n are contained. Clearly, on A, $p_n = 1$ for all n. This proves that A is contained in the poles of p. We have proved

Theorem 8.3.2. *A set is polar if and only if it has capacity zero, which holds if and only if it is quasi null. In particular, the set of irregular points in a compact set (being quasi null by* Proposition 8.1.1) *is polar.*

Application in measure theory. Using the capacity theorem, we now show that hitting time to an analytic set is measurable provided the fields are complete. Precise formulations follow.

Let us first discuss some application of the capacity theorem to measure theory. Let Y be an analytic space and P a Borel probability measure on Y, i.e., P is a probability measure defined on all Borel subsets of Y. For an arbitrary subset A contained in Y, define

$$C(A) = \inf P(U), \qquad (8.3.1)$$

where the infimum is over all open sets containing A. C has properties (1) and (3) of the definition of Choquet capacity given before Theorem 8.2.4. It also has property (2) because P being a measure (iii) of Theorem 8.2.2 is clear for open sets A and B and hence also for arbitrary sets:

$$C(A \cup B) + C(A \cap B) \leq C(A) + C(B). \qquad (8.3.2)$$

The argument leading from (iii) of Theorem 8.2.2 to (iv) of Theorem 8.2.2 also applies here. Thus, C defined by (8.3.1) is a Choquet capacity on all subsets of yY.

Now we show that C agrees with P on all Borel sets. To this end, let us show that C is additive on disjoint analytic sets:

$$C(A \cup B) = C(A) + C(B), \quad A, B \text{ disjoint analytic}. \qquad (8.3.3)$$

From (8.3.2), the left side of (8.3.3) is less or equal to the right side. It is sufficient, by Theorem 8.2.4, to prove reverse inequality for disjoint compact sets A and B. This is immediate from the following three observations: C agrees with P on open sets; P is additive on disjoint open sets; given any open set U containing the compact set $A \cup B$, we can find disjoint open neighborhoods of A and B with union contained in U. Thus, (8.3.3) is established. Finally, let A be any Borel set and B its complement. C clearly dominates P on Borel sets. The equality

$$1 = P(A) + P(B) \leq C(A) + C(B) = C(Y) = 1$$

shows that C and P agree on Borel sets.

Observe also that for an analytic set A, there is an increasing sequence K_n of compact subsets of A and a decreasing sequence V_n of open sets containing A such that $C(A)$ agrees with both $P(\cup_n K_n)$ and $P(\cap_n V_n)$. In other words, A is "measurable" relative to P.

We collect all this in the following:

Theorem 8.3.3. *Every Borel measure on an analytic space Y is regular, i.e., the measure of every Borel set is approximable from within by the measures of compact subsets. Every analytic subset of Y is universally measurable, i.e., measurable relative to every Borel probability measure on Y.*

Now let us look at hitting times. Fix $t > 0$ and let Z denote the separable Banach space of continuous functions on $[0, t]$ into $\mathbb{R}^d$. $Y = (0, t] \times Z$ is a Polish space and the map X,

$$X(s, w) = w(s) \in \mathbb{R}^d$$

is continuous on Y. The inverse image of any analytic subset of $\mathbb{R}^d$ is analytic in Y: Indeed, the graph of X, grX, is a closed subset of $Y \times \mathbb{R}^d$ because X is continuous. If A is analytic subset of $\mathbb{R}^d$, $X^{-1}(A)$ is simply the Y-projection of the analytic set $(grX) \cap (Y \times A)$. Projection is continuous, and continuous image of an analytic set (being the image under a composite map of $\mathcal{N}$) is analytic. The set

$$\{w \in W : w(s) \in A \text{ for some } 0 < s \leq t\} \qquad (8.3.4)$$

being the Z-projection of the set $X^{-1}(A)$ is analytic and therefore universally measurable as a subset of Z by Theorem 8.3.3. (See also the Exercises to Section 8.2.)

Now the Borel field of Z is the Borel field generated by open balls. By continuity the norm $\|w - w_0\|$ is the supremum, over the rationals r in $[0, t]$, of $|w(r) - w_0(r)|$. The norm is thus measurable relative to Borel field $\mathbb{B}$ generated by the coordinate maps: $w \mapsto w(s)$, $0 \le s \le t$. The Borel sets in Z are thus simply the elements of $\mathbb{B}$. The set (8.3.4) is thus measurable relative to every probability measure on $\mathbb{B}$. The relation between this Borel field $\mathbb{B}$ and the stopped Borel field B_t introduced in Section 3.1 is clear. Now the reader should have no difficulty in the following:

Theorem 8.3.4. *Let A be an analytic subset of $\mathbb{R}^d$ and H its hitting time,*

$$H = \inf\{s : s > 0,\ X_s \in A\},$$

the infimum over an empty set being defined ∞. Then the set $\{H \le t\}$ is measurable relative to every probability measure on B_t.

Remark. The hitting time to an analytic set can be approximated by the hitting times to compact subsets. Briefly, the details are as follows. Let P be any probability measure on B_∞ = the Borel field introduced in Section 3.1. Given $t > 0$, we can regard P as a probability measure on the stopped Borel field B_t which can be identified with the Borel field of Z. Let π be the projection of Y onto Z:

$$\pi(s, w) = w.$$

For any subset $B \subset Y$, let

$$C(B) = P^*(\pi(B)),$$

where P^* is the outer measure corresponding to P. Since π is continuous, C is seen to be a Choquet capacity on all subsets of Y — note that if B_n decreases strongly to B then $\pi(B_n)$ decreases strongly to $\pi(B)$.

Now let A be an analytic subset of $\mathbb{R}^d$ and $B = X^{-1}(A)$. B is analytic in Y. $C(B)$ is the supremum of $C(F)$ with F compact in B. In particular, X being continuous: $C(B) = \sup C(X^{-1}(K))$, K compact in A.

Calling the set in (8.3.4) $R(t, A)$, what has shown is this: For each $t > 0$, there is an increasing sequence $K_j(t)$ of compact subsets of A such that $P(R(t, K_j(t)))$ increases to $P(R(t, A))$. Enumerate the non-negative rationals $\{r_n\}$. $K_n = K_n(r_1) \cup \cdots \cup K_n(r_n)$ gives an increasing sequence of compact subsets of A for which, for every rational r, $P(R(r, K_n))$ increases to $P(R(t, A))$. Now it is easy to show that the hitting times to K_n decrease to the hitting time to A, P-almost everywhere.

8.4 Balayage

Physically, the problem of balayage is the following: Given a compact set F and a spatial distribution of charges m, can we find a distribution of charges n on F such that the potential on F is unaltered?

We have seen in Section 8.1 that given a measure m and a compact set F we can find a measure n with support F such that the potentials of m and n agree on F except on a subset of capacity zero (c.f. Theorem 8.3.2). The Maria–Frostman maximum principle of Section 8.1 shows that any non-negative superharmonic function which dominates the potential of m quasi everywhere on F (i.e., except for a set of capacity zero), dominates the potential of n everywhere. Since the potential of n is also superharmonic, we can state this as follows: the potential of $n = $ infimum of all non-negative superharmonic functions which dominate, quasi everywhere on F, the potential of m.

We now describe the well-known balayage technique. As before W will denote a Green domain $\subset \mathbb{R}^d$, $d \geq 2$. For a real valued function f, its lower semicontinuous regularization, denoted $\hat{f}$, is defined by

$$\hat{f}(x) = \liminf_{y \to x} f(y).$$

$\hat{f}$ is the largest lower semicontinuous function less or equal to f. Let $E \subset W$ and φ a non-negative function on E. The *reduit* or *reduced function* R_φ^E is defined to be the infimum of all non-negative hyperharmonic functions (on W) which dominate φ on E:

$$R_\varphi^E = \inf\{u : u \geq 0 \text{ hyperharmonic}, u \geq \varphi \text{ on } E\}.$$

The lower semicontinuous regularization of R_φ^E, denoted $\hat{R}_\varphi^E$, is called the *balayage of φ relative to E*. This is standard notation. However, we some times write $R(E, \varphi)$ and $\hat{R}(E, \varphi)$ respectively for reduit and balayage.

Now we need to generalize Lemma 8.1.4 to an arbitrary family of excessive functions. To this end, we need the following lemma.

Lemma 8.4.1. *Let $\{f_i : i \in I\}$ be a family of extended real valued functions on W. For $J \subset I$, put $f_J(x) = \inf_{i \in J} f_i(x)$, $x \in W$. Then there exists a countable subset $I_0 \subset I$ such that $\hat{f}_{I_0} = \hat{f}_I$.*

Proof. Replacing f_i by $\arctan f_i$ if needed, assume that f_i are uniformly bounded. Let $\{U_j\}$ be a countable base for the topology on W. Let $x_j \in U_j$

satisfy
$$f_I(x_j) < \inf_{x \in U_j} f_I(x) + \frac{1}{2j}$$

and f_{i_j} be such that $f_{i_j}(x_j) < f_I(x_j) + \frac{1}{2j}$. Then

$$\inf_{x \in U_j} f_{i_j}(x) \leq f_{i_j}(x_j) < \inf_{x \in U_j} f_I(x) + \frac{1}{j}.$$

Let I_0 be the set $\{i_1, i_2, \ldots\}$. The last inequality implies that, for all j,

$$\inf_{x \in U_j} f_{I_0}(x) < \inf_{x \in U_j} f_I(x) + \frac{1}{j},$$

i.e., that $\hat{f}_{I_0} \leq \hat{f}_I$. □

Now the following theorem is immediate from the above Lemmas 8.4.1, 8.1.4, and Theorem 8.3.2:

Theorem 8.4.2. Let $\{s_i : i \in I\}$ be a family of excessive functions and $s = \inf s_i$. Then $\hat{s}$ is excessive and the set $\{\hat{s} < s\}$ is polar.

From the above theorem, the reduit and balayage differ at most on a polar set. And balayage of φ is superharmonic.

We can define balayage directly as follows: Suppose $F \subset E$ is polar, u hyperharmonic and $\geq \varphi$ on $E \backslash F$. $x_0 \notin F$. By Proposition 8.3.1, there is a potential p with $p(x_0) \leq 1$ and $p = \infty$ on F. $u + \epsilon p \geq \varphi$ on E and hence $u + \epsilon p \geq \hat{R}(E, \varphi)$. Letting ϵ tend to zero, we see $u(x_0) \geq \hat{R}(E, \varphi)(x_0)$, i.e., $u \geq R(E, \varphi)$ except perhaps on F. E being polar it has measure zero: $u \geq \hat{R}(E, \varphi)$ everywhere. Thus, we can also define (since $\hat{R}(E, \varphi) \geq \varphi$ on E except for a polar set)

$$\hat{R}(E, \varphi) = \inf\{u : u \text{ hyperharmonic}, u \geq \varphi \text{ quasi everywhere on } E\}.$$
(8.4.1)

When $E = W$, reduit and balayage will simply be denoted by R_φ and $\hat{R}_\varphi$, respectively. Clearly, $R_\varphi^E = R_{1_E \varphi}$.

The following remark is useful in the proof of Lemma 8.4.3.

Remark. For any x, $R_\varphi(x) > \varphi(x)$ implies $\hat{R}_\varphi(x) = R_\varphi(x)$. This is the same as saying $\hat{R}_\varphi(x) < R_\varphi(x)$ implies $R_\varphi(x) = \varphi(x)$. Indeed let $F = \{\hat{R}_\varphi < R_\varphi\}$. F is polar. Let $x_0 \in F$. If $\varphi(x_0) = \infty$, there is nothing to show. If $\varphi(x_0) < \infty$, let p be a potential, $p(x_0) \leq 1$ and $p = \infty$ on $F\setminus\{x_0\}$. $\hat{R}_\varphi + \epsilon p \geq \varphi$ everywhere for any ϵ such that $\hat{R}_\varphi(x_0) + \epsilon p(x_0) \geq \varphi(x_0)$. Were $\hat{R}_\varphi(x_0) \geq \varphi(x_0)$ we could let ϵ tend to zero to get $\hat{R}_\varphi(x_0) = R_\varphi(x_0)$, i.e., $x_0 \notin F$. We must therefore have $\hat{R}_\varphi(x_0) < \varphi(x_0)$. But in this case for suitable ϵ, $\hat{R}_\varphi(x_0) + \epsilon p(x_0) = \varphi(x_0)$, i.e., $R_\varphi(x_0) = \varphi(x_0)$.

Lemma 8.4.3. Let $0 \leq \varphi_n$ increase to φ. Then

$$\sup_n R_{\varphi_n} = R_\varphi, \quad \sup_n \hat{R}_{\varphi_n} = \hat{R}_\varphi. \tag{8.4.2}$$

Proof. Since a countable union of polar sets is polar, the second equality in (8.4.2) follows from (8.4.1). Using the second equality in (8.4.2) and the above remark, $\hat{R}_\varphi(x_0) < R_\varphi(x_0)$ implies $R_\varphi(x_0) = \varphi(x_0)$, i.e., $\hat{R}_{\varphi_n}(x_0) < \varphi_n(x_0)$ from some n on. Again by the same remark, this in turn replies that $R_{\varphi_n}(x_0) = \varphi_n(x_0)$ for those n for which $\hat{R}_{\varphi_n}(x_0) < \varphi_n(x_0)$. The first part of (8.4.2) thus follows for those x for which $\hat{R}_\varphi(x) < R_\varphi(x)$. For other x's, this is a consequence of the second part of (8.4.2). □

Lemma 8.4.4. Let A_n decrease to A strongly, i.e., each open set containing A contains some A_n. If $\varphi > 0$ is finite and upper semicontinuous, then

$$\lim R(A_n, \varphi) = R(A, \varphi). \tag{8.4.3}$$

Proof. If s is excessive and $\geq \varphi$ on A, then for any $\epsilon > 0$, $s + \epsilon > \varphi$ in an open set containing A. Such an open set will contain some A_n. And for such n, $s + \epsilon > R(A_n, \varphi)$, proving that $\epsilon + R(A, \varphi) \geq \inf_n R(A_n, \varphi)$. The reverse inequality being clear (8.4.3) follows. □

The above two lemmas assert that for at finite upper semicontinuous φ, for every x, $R(A, \varphi)(x)$ is a Choquet capacity on all subsets of our Green domain W.

Theorem 8.4.5. Let $\varphi \geq 0$ be finite and upper semicontinuous. Then

$$R(E, \varphi) = \inf R(D, \varphi), \tag{8.4.4}$$

where the infimum is over all open sets D containing E. If E is analytic,

$$R(E, \varphi) = \sup R(A, \varphi), \tag{8.4.5}$$

where the supremum is over all compact subsets of E and there exists an increasing sequence A_n of compact subsets of E such that

$$\hat{R}(E, \varphi) = \sup_n \hat{R}(A_n, \varphi). \tag{8.4.6}$$

Proof. If s is excessive dominates φ on E, then for every $\epsilon > 0$, $s + \epsilon$ dominates φ on an open set containing E and (8.4.4) follows. Equation (8.4.5) is a consequence of the capacity theorem.

To prove (8.4.6), note that the family $\{\hat{R}(A, \varphi) : A \text{ compact } \subset E\}$ is filtering to the right, i.e., any two members are both dominated by a third. For any compact set $K \subset W$,

$$\sup_{\text{compact } A \subset E} \int_K \hat{R}(A, \varphi)(x)\, dx = \int_K \hat{R}(E, \varphi)(x)\, dx \qquad (8.4.7)$$

because the balayage is almost everywhere equal to the reduit and we are using (8.4.5). Now let K_n be compacts increasing to W and choose compact subsets $A_n \subset E$ so that the integrals in (8.4.7) differ by at most $\frac{1}{n}$. If $B_n = A_1 \cup \cdots \cup A_n$, $\sup_n \hat{R}(B_n, \varphi)$ has the same integral as $\hat{R}(E, \varphi)$ on all compact subsets of W and hence they are identical. □

Interpretation. For a compact set F, as is clear from (8.4.1), $\hat{R}(F, 1)$ is simply the hitting potential of F. The same holds for an analytic set. Indeed, if H is the hitting time to the analytic set A, using (8.4.4), $\mathbb{P}.[H < R] \leq \hat{R}(A, 1)$, $R = $ exit time from W. The reverse inequality follows from (8.4.6).

Thinness. A set A is called *thin at a point* x if $x \notin \overline{A}$ or if $x \in \overline{A}$ and there is a potential p on W such that

$$\liminf_{A \setminus \{x\} \ni y \to x} p(y) > p(x). \qquad (8.4.8)$$

Thinness can also be defined as follows: A is thin at $x \in \overline{A}$ if and only if there is a neighborhood U of x such that for $B = A \cap U$,

$$\hat{R}(B, 1)(x) < 1. \qquad (8.4.9)$$

Indeed, if p satisfies (8.4.8), we can find a neighborhood U of x such that the potential $q = p|p(x)$ is larger than or equal to b in $U \cap A = B$, where b is a number larger than 1 and $q(x) = 1$. The potential $b^{-1}q$ dominates 1 on B and is strictly less than 1 at x. Equation (8.4.9) must therefore be true. Conversely, suppose (8.4.9) holds. If $E = B \setminus \{x\}$, then $\hat{R}(E, 1) = \hat{R}(B, 1)$. (From (8.4.1), it should be clear that polar sets can be added or subtracted without affecting balayage.) Since x is not in E, we have $\hat{R}(E, 1)(x) = R(E, 1)(x)$. This is easy to see, c.f. Exercise 8.4.3. And the first term is less than 1. Now use the definition of $R(E, 1)$ to find a potential p, dominating 1 on E such that $p(x) < 1$. This p satisfies (8.4.8).

Probabilistically, thinness is explained as follows. Let A be thin at $x \in \overline{A}$ and p a potential satisfying (8.4.8). If α is a number between the quantities

Potential Theory, Capacity, Boundaries, Dirichlet Spaces, and Applications   213

in (8.4.8), the open set $D = \{p > \alpha\}$ contains $A \setminus \{x\}$ and $x \in \partial D$. p is excessive and $p(x) < \alpha$. The hitting time to D, starting at x, cannot therefore be zero. Thus the Brownian path, starting at x, remains in the complement of A for a positive time. It is clear that D *itself is thin at* x.

If F is compact, it is clear from (8.4.9) that F is thin at a point $x \in F$ if and only if x is irregular for F. We know that the set of irregular points in a compact set if polar (Theorem 8.3.2). More generally, we have the following:

Proposition 8.4.6. *For any set A, the set of points $x \in A$ at which A is thin is polar. In particular, a set is polar if and only if it is thin at each of its points.*

Proof. Since a countable union of polar sets is polar, by (8.4.9) we need only show the following: For each $a > 0$, the set E of $x \in A$ for which $\hat{R}(V_x, 1)(x) < 1$, where $V_x = B(x, 3a) \cap A$, is polar. The open cover of E consisting of ball of radius a around each point of E has a countable sub-cover: There is a countable set $I \subset E$ such that every $y \in E$ belongs to $U_x = B(x, a) \cap A$ for some $x \in I$. But if $x \in I$ and $y \in U_x$, then $B(y, 3a) \supset B(x, a)$ so that by the definition of E, $\hat{R}(U_x, 1)(y) < 1$ for all $y \in U_x \cap E$, i.e., $U_x \cap E$ is polar. Thus E is a countable union of polar sets.

That a polar set is thin at each of its points is contained in Proposition 8.3.1. □

Exercises

8.4.1. Show that $R(E, \varphi)$ is Lebesgue measurable.

8.4.2. Show that $R(E, \varphi)$ is harmonic off $\overline{E}$ unless it is identically infinite.
Hint: If $B(a, r)$ is disjoint from $\overline{E}$, $T =$ exit time from $B(a, r)$ and s is excessive and dominates φ on E, then $\mathbb{E}.[s(X_T) : T < R]$ does the same. $R =$ exit time from W.

8.4.3. $R(E, \varphi) = \hat{R}(E, \varphi)$ off E. If φ is continuous, equality holds also in the interior of E.
Hint: For the first use the Remark before Lemma 8.4.1. For the second note that balayage is the largest lower semicontinuous function less or equal to the reduit.

8.4.4. If u and v are continuous and excessive,
$$R(E, u) + R(E, v) = R(E, u + v).$$

Hint: By (8.4.4), one may assume $E = D$ is open. If H is the hitting time to D, $R(D, u) = \mathbb{E}. [u(X_H) : H < R]$. Note that by Exercise 8.4.3, balayage and reduit are the same for open sets.

8.4.5. If u is continuous and excessive, then

$$R(A \cup B, u) + R(A \cap B, u) \leq R(A, u) + R(B, u).$$

This is strong sub-additivity.

Hint: By (8.4.4), sufficient to show this assuming A and B are open. If H_1 and H_2 are hitting times to A and B, then $H = H_1 \wedge H_2$ is the hitting time to $A \cup B$ and $H_1 \vee H_2 = S$ is less or equal to the hitting time I to $A \cap B$. Sum on the right side is simply $\mathbb{E}. [u(X_H) : H < R] + \mathbb{E}. [u(X_S) : S < R]$. The first term is just $R(A \cup B, u)$. The second term dominates $R(A \cap B, u) = \mathbb{E}. [u(X_I) : I < R]$ because u is excessive and $I \geq S$.

8.4.6. E is polar if and only if $\hat{R}(E, 1) \equiv 0$.

Hint: Suppose $\hat{R}(E, 1) \equiv 0$ and $x_0 \notin E$. There exists excessive functions s_n such that $s_n(x_0) < 2^{-n}$ and $s_n \geq 1$ on E.

8.4.7. If E is relatively compact, $\hat{R}(E, \varphi)$ is a potential unless it is identically infinite.

Hint: If u is excessive, not identically infinite, dominates φ on E, D relatively compact open and contains E, then $p(\cdot) = \mathbb{E}. [u(X_H) : H < R]$ is a potential and equals u on E. $R = $ exit time from the Green domain W.

8.4.8. A is thin at $x \in \overline{A}$ if and only if there is a super harmonic function s such that $\liminf_{A \ni y \to x} s(y) = s(x)$.

Hint: Consider the potential of the restriction of the Riesz measure of s to a relatively compact neighborhood of x.

8.4.9. Let A and B be thin at x. Then $A \cup B$ is thin at x.

8.5 Dirichlet Spaces

In the following, W is a Green domain in $\mathbb{R}^2$, $d \geq 2$. A complex Radon measure m will be said to have finite energy if $|m| = $ total variation measure of m has finite energy. If m and n are complex Radon measures of finite energy, the expression

$$(m, n) = \int G(x, y) \, m(dx) \, \overline{n}(dy) \tag{8.5.1}$$

clearly makes sense; $\bar{n}$ is the complex conjugate of n. And from the energy principle $\|m\|^2 = (m,m)$ is positive unless $m = 0$. The space ξ of all complex Radon measures of finite energy is a pre-Hilbert space with the inner product (8.5.1). For each $x \in \xi$, the function Gm, which is defined except for a set of capacity zero, will be called the potential of m. Since $|Gm| \leq G\,|m|$, if h is complex harmonic and $|h| \leq |Gm|$, then $h = 0$; indeed if u is the real part of h, then $u < |Gm|$ so that $u \leq 0$ and for the same reason $-u \leq 0$. The name complex potential is thus not unjustified. The complex potential Gm completely determines m by

$$\int \Delta\varphi Gm = -A_d \int \varphi\, dm \qquad (8.5.2)$$

for every C^∞-complex function φ with compact support. m is thus the *Riesz measure* of Gm. If f and g are complex potentials with Riesz measures m and n,

$$(f,g) = (m,n) = \int f\, d\bar{n} = \int \bar{g}\, dm \qquad (8.5.3)$$

defines an inner product. (f_i) is a Cauchy sequence if and only if the corresponding Riesz measures (m_i) is a Cauchy sequence. Now the energy of a complex Radon measure is the sum of the energies of its real and imaginary parts. It follows that the real and imaginary parts of (f_i) are themselves Cauchy sequences. An argument as in the proof of Theorem 8.1.6 shows that there is a function f such that f_i tends to f in $L^1(dn)$ for every $n \in \xi$. Indeed, a candidate for f can be found as follows: For a subsequence, we call (l_i), of (m_i), $\sum \|l_{i+1} - l_i\| < \infty$. If $u_j + iv_j = Gl_j$, $f = \liminf u_j + i\liminf v_j$ is such a function. Obviously, any two such functions f can differ at most on a set of capacity zero. Thus the completion of the pre-Hilbert space of complex potentials can be identified with a set of functions; two of these functions considered the same if they agree quasi everywhere. *These are the so-called BLD functions.*

There is another way of defining the class of BLD functions. The claim is that the completion of the pre-Hilbert space of all complex valued C^∞-functions with compact supports in W, provided with the inner product

$$(\varphi, \psi) = \int (\operatorname{grad}\varphi, \overline{\operatorname{grad}\psi}) \qquad (8.5.4)$$

is simply the class of BLD functions.

Every complex C^∞-function φ with compact support is a complex potential whose Riesz measure is (up to a constant) $\Delta\varphi$. The inner product in (8.5.4) is up to a constant $\int \bar{\psi} \Delta \varphi$ as is seen by partial integration. Thus the inner product in (8.5.4) is up to a constant the inner product of complex potentials given in (8.5.3). The claim will therefore be shown by the following Lemma 8.5.1.

Let us collect here a few elementary properties of BLD functions. This will be useful in the proof of Lemma 8.5.1. All these are immediate from the definition:

1. If f is a BLD function and g a complex potential with Riesz measure n, then $f \in L^1(dn)$ and the scalar product

$$(f,g) = \int f \, d\bar{n} \quad \text{and} \quad \left| \int f \, d\bar{n} \right| \leq \|f\| \|n\|, \quad n \in \xi. \tag{8.5.5}$$

2. The restriction of the Lebesgue measure to any compact set being of finite energy implies that every BLD function is locally integrable.

3. If $m_i \in \xi$ converges to $m \in \xi$, then $\int f \, dm_i$ converges to $\int f \, dm$ for every BLD function f.

Lemma 8.5.1. *The set of all C^∞-functions with compact supports is dense in the space of complex potentials.*

Proof. Step 1: The set of $m \in \xi$ of compact support is dense in ξ; the general case follows from the case of positive measures. For $m \in \xi_+$, this is obvious from: $\int_F \int_F G(x,y) \, m(dx) \, m(dy)$ increases to (m,m) as the compact set F increases to W.

Step 2: The set of m with compact support and smooth density is dense in ξ. Suppose $m \in \xi_+$ has compact support and $p = Gm$. Let $0 \leq \varphi_i$ be C^∞ and radial (i.e., depends only no distance) and have support in $B(0, 1/i) =$ the ball with center zero and radius $1/i$, and $\int \varphi_i = 1$. If p_i is the potential of $m_i = m * \varphi_i$,

$$p_i = \int m(dy) \int G(\cdot, z) \varphi_i(z-y) \, dz \leq p \tag{8.5.6}$$

for all large i and tends to p because the inner integral is less or equal to and tends to $G(\cdot, y)$ on support m as soon as $B(y, 1/i) \subset W$ for all y in support m (see the discussion on approximation of superharmonic functions by smooth ones in the beginning of Section 6.1). Now

$$0 \leq \|m - m_i\|^2 = \int p \, dm - 2 \int p_i \, dm + \int p_i \, dm_i \leq \int p \, dm - \int p_i \, dm \tag{8.5.7}$$

because $\int p_i \, dm_i \leq \int p \, dm_i = \int p_i \, dm$. Since $p_i \leq p$ and tends to p, the last term in (8.5.7) tends to zero. thus m_i tends to m in the energy norm.

Potential Theory, Capacity, Boundaries, Dirichlet Spaces, and Applications   217

Step 3: If the claim in the Lemma were false, there would exist a BLD function f such that for every C^∞-function φ with compact support $\int f \Delta \varphi = 0$. By Lemma 6.3.1, there is a (complex) harmonic function h such that $f = h$ almost everywhere. Now let $m \in \xi_+$ have compact support. If φ_i are as in Step 2, $m_i = m * \varphi_i$ tends to m in ξ_+ and m_i tends to m vaguely.

For all large enough i, the supports of m_i are contained in a fixed compact subset of W. Also m_i has smooth density, and

$$\left|\int h\,dm\right| = \lim \left|\int h\,dm_i\right| = \lim \left|\int f\,dm_i\right| = \left|\int f\,dm\right|. \qquad (8.5.8)$$

Let D be open relatively compact and $a \in \partial D$. The harmonic measure at a relative to D has small energy if D is large (see the proof of Lemma 8.1.5). h being harmonic the first term in (8.5.8) is $|h(a)|$ if m is the harmonic measure at a relative to D. And the last term in (8.5.8) is small if D is large. Thus $h \equiv 0$. But then the last term in (8.5.8) is zero for all $m \in \xi$ with compact support and hence for all $m \in \xi$. Thus $f = 0$ quasi everywhere. See also Exercise 8.5.2.

Lemma 8.5.1 easily implies the following result about continuity properties of BLD functions: *If f is BLD and $\epsilon > 0$ there is an open set of capacity smaller than ϵ such that the restriction of f to the complement of this open set is continuous.* This is Exercise 8.5.3.

For the following corollary, we need a definition. A locally integrable function g in W is said to have a *generalized gradient*, denoted $\operatorname{grad} g = (g_1, \ldots, g_d)$, if for every C^∞-function φ with compact support (in W),

$$\int g \frac{\partial \varphi}{\partial \varphi_i} = -\int g_i \varphi, \quad 1 \le i \le d. \qquad (8.5.9)$$

We say that $\operatorname{grad} g \in L^2$ if $g_i \in L^2$ for $1 \le i \le d$. With this definition, we have the following:

Corollary 8.5.2. *If f is BLD then the generalized $\operatorname{grad} f$ exists and $\|f\|^2 = \int |\operatorname{grad} f|^2$.*

Conversely, if g has a generalized gradient in L^2, then there exists an f in BLD and h harmonic such that $g = f + h$ almost everywhere.

Proof. By Lemma 8.5.1, if f is BLD, then there exist C^∞-functions φ_i with compact supports such that $\|\varphi_i - f\|$ tends to zero. In particular, φ_i is a Cauchy sequence. As we have remarked, the (energy) norm $\|\varphi_i - \varphi_j\|$ is, up to a constant, the L^2-norm of $\operatorname{grad}(\varphi_i - \varphi_j)$. So $\operatorname{grad}\varphi_i$ converges in L^2 (and in particular locally in L^1) to, say, $(f_1, \ldots, f_d)$. If ψ is C^∞ with compact support (recall that f is locally integrable and that φ_i converges to f in $L^1(dn)$ for every $n \in \xi_+$, in particular locally in L^1), then

$$\int f \frac{\partial \psi}{\partial x_1} = \lim \int \varphi_i \frac{\partial \psi}{\partial x_1} = -\lim \int \frac{\partial \varphi_i}{\partial x_1} \psi = -\int f_1 \psi,$$

same conclusion holding for other partials. Thus, $(f_1, \ldots, f_d)$ is the generalized gradient of f.

Conversely, suppose g has generalized gradient $\operatorname{grad} g = (g_1, \ldots, g_d) \in L^2(W)$. For every φ which is C^∞ with compact support,

$$\int g \Delta \varphi = -\int (\operatorname{grad} g, \operatorname{grad} \varphi) \tag{8.5.10}$$

as is clear from (8.5.9). Since $\operatorname{grad} g$ is in L^2, and the L^2-norm of $\operatorname{grad} \varphi$ is up to a constant the energy norm of $\Delta \varphi$, (8.5.10) shows by Lemma 8.5.1 that g defines a continuous linear functional on the Hilbert space of BLD functions. So there is an f in BLD which determines the same linear functional, say g: $\int f \Delta \varphi = \int g \Delta \varphi$, for every C^∞-function φ with compact support. g being locally integrable, $g - f$ is thus almost everywhere equal to a harmonic function h. □

Remark. A BLD function f is absolutely continuous on almost all lines parallel to the coordinate axes and the gradient in the usual sense exists almost everywhere. See Exercise 8.5.5. In any case, the gradient of the potential of a positive measure m exists almost everywhere (Exercise 6.3.1) and is locally integrable. Thus the energy norm of a positive measure is up to a constant the L^2-norm of its gradient. See Exercise 8.5.4. The same of course is true for any $m \in \xi$.

The space of BLD functions has one more important property. To describe this we need a definition. A function B on the complex plane into itself is called a *normal contraction* if $B(0) = 0$ and

$$|B(x) - B(y)| \leq |x - y|, \quad x, y \text{ complex}. \tag{8.5.11}$$

Theorem 8.5.3. *Normal contractions operate on the space of BLD functions: If f is BLD, so is Bf for any normal contraction B and $\|Bf\| \leq \|f\|$.*

Proof. A Lipschitz function (i.e., a function f which satisfies $|f(x) - f(y)| < M |x - y|$) is absolutely continuous on every line. Therefore, its restriction to any line is differentiable almost everywhere with a bounded derivative; see Rudin [47, p. 165]. In particular, if a Lipschitz function f has compact support, it has a generalized gradient which is in L^2. If $0 \le \varphi_i$ are C^∞ with small supports around the origin and $\int \varphi_i = 1$, then $f_i = f * \varphi_i$ will be C^∞ with compact support (in W); it converges to f and $\operatorname{grad} f_i = \operatorname{grad} f * \varphi_i$ converges in L^2 to $\operatorname{grad} f$. Hence f is BLD.

Next, if φ is C^∞ with compact support and B a normal contraction, then the support of $B\varphi$ is contained in the support φ and, because φ is clearly Lipschitz, (8.5.11) shows that $B\varphi$ is also Lipschitz. Also $|B\varphi(x) - B\varphi(y)| \le |\varphi(x) - \varphi(y)|$ shows that $|\operatorname{grad} B\varphi|^2 \le |\operatorname{grad} \varphi|^2$. Thus, $B\varphi$ is BLD and its BLD norm does not exceed the BLD norm of φ. [Recall that the BLD norm of a BLD function is up to a constant the L^2-norm of its (generalized) gradient.]

Finally, let g be a BLD function. Then there is a sequence g_i of C^∞-functions with compact supports such that g_i tends to g in $L^1(dn)$ for every $n \in \xi_+$, and hence also for all $n \in \xi$. By (8.5.11), Bg_i tends to Bg in $L^1(dn)$ for all $n \in \xi$. And

$$\left| \int Bg\, dn \right| = \lim \left| \int Bg_i\, dn \right| \le \lim \|g_i\| \|n\| = \|g\| \|n\|,$$

where $\|\cdot\|$ denotes the energy or BLD norm. So there is an $f \in$ BLD such that $\int Bg\, dn = \int f\, dn$ for all $n \in \xi$. But then $Bg = f$ quasi everywhere. That proves the theorem. □

Let $\mathbb{D}$ denote a class of functions on W having the following four properties. That the space of BLD functions has these properties has been established:

1. $\mathbb{D}$ is a Hilbert space, with norm $\|\cdot\|$ and scalar product $(\cdot,\cdot)$, of locally integrable functions on W.
2. For any compact set F,

$$\int_F |f| \le A(F) \|f\|, \quad f \in \mathbb{D},$$

where $A(F)$ is a constant depending only on F. To see this for BLD, consider the Borel function g defined on F by $g = |f|/f$ if $f \ne 0$ and zero otherwise; $g = 0$ off F. The scalar product (f, Gg) is then simply the integral of $|f|$ on F. And the energy norm of g is less or equal to the energy norm of the indicator of F. This property for BLD is also an immediate consequence of Property 4.

3. If $K(W)$ = set of continuous functions with compact support, then $K(W) \cap \mathbb{D}$ is dense both in $\mathbb{D}$ and in $K(W)$. The density in $K(W)$ is understood as follows: For any $f \in K(W)$ and any neighborhood U of the support of f and any $a > 0$, there is a g in $K(W) \cap \mathbb{D}$ with support in U satisfying $\sup|f - g| < a$.
4. For any normal contraction B and $f \in \mathbb{D}$, $Bf \in \mathbb{D}$ and $\|Bf\| \le \|f\|$.

The above four properties characterize the so called *Dirichlet spaces*. We now give a rough sketch of the possibilities in the direction of a kernel-free potential theory implicit in the above properties. The reader is invited to elaborate on these ideas himself.

If $f \in \mathbb{D}$, f and $\overline{f}$ are contractions of each other; so $\overline{f} \in \mathbb{D}$ and $\|f\| = \|\overline{f}\|$. $z \mapsto |z|$ is a normal contraction so $f \in \mathbb{D}$ implies $|f| \in \mathbb{D}$.

$f \in \mathbb{D}$ is called a *pure potential* if there is a non-negative measure m such that

$$(f, \varphi) = \int \overline{\varphi}\, dm \qquad (8.5.12)$$

for every $\varphi \in K(W) \cap \mathbb{D}$. The reader is invited to check that in the case of BLD functions f is a pure potential if and only if $f = Gm$ almost everywhere. Note that the four properties by themselves do not permit us to distinguish between functions which are equal almost everywhere. A purely geometric characterization of a pure potential is given by the following important proposition.

Proposition 8.5.4. $f \in \mathbb{D}$ *is a pure potential if and only if*

$$\|g + f\| \ge \|f\| \qquad (8.5.13)$$

whenever $g \in \mathbb{D}$ *and* $\Re g \ge 0$. *Equation* (8.5.13) *is equivalent to*

$$\Re(f, g) \ge 0 \quad \text{provided } \Re g \ge 0. \qquad (8.5.14)$$

Proof. To see (8.5.13) and (8.5.14) are equivalent, take $a \cdot g$ in stead of g in (8.5.13) where $a > 0$ and let a tend to zero. If f is a pure potential, (8.5.14) is a consequence of (8.5.12) for $g \in K(W) \cap \mathbb{D}$ and the positivity of m; the general case then follows by continuity.

Suppose now that f satisfies (8.5.13). The closed convex set

$$F = \{f + g : g \in \mathbb{D}, \Re g \ge 0\}$$

contains a unique element of least norm, which by (8.5.13) must be f. Since $|f| \in F$ and its norm cannot exceed that of f, we must have $f = |f|$.

If $\varphi \in K(W) \cap \mathbb{D}$, so is $|\varphi|$. So there are enough positive functions in this intersection. The map $\varphi \mapsto (f, \varphi)$ is a positive linear functional on $K(W) \cap \mathbb{D}$ by (8.5.14) and must be given by a unique measure m. □

Corollary 8.5.5. *If u and v are pure potentials, so is $u \wedge v$.*

Proof. Since $f \in \mathbb{D}$ implies $|f| \in \mathbb{D}$, $w = u \wedge v = \frac{1}{2}[u + v - |u - v|] \in \mathbb{D}$. The closed convex set $F = \{f : \Re(f - w) \geq 0\}$ contains a unique element, say h, of least norm. h must be a pure potential because for any g with $\Re g \geq 0$, $h + g \in F$ and so $\|h + g\| \geq \|h\|$. Clearly, $u \wedge h \in F$ and

$$4\|u \wedge h\|^2 = \|u + h\|^2 + \||u - h|\|^2 - 2(u + h, |u - h|)$$
$$\leq \|u + h\|^2 + \|u - h\|^2 - 2(u + h, u - h) = 4\|h\|^2$$

because u being a pure potential, $(u, |u - h| - (u - h)) \geq 0$ by (8.5.14) and similarly for h. By uniqueness, $u \wedge h = h$, i.e., $h \leq u$. For the same reason, $h \leq v$, i.e., $h \leq u \wedge v = w$. □

Each bounded non-negative measurable function f with compact support determines a pure potential Gf (just notation) by the prescription

$$(Gf, v) = \int f\bar{v}, \quad v \in \mathbb{D}.$$

Indeed, by Property 2, the map $v \mapsto \int f\bar{v}$ is a continuous linear functional on $\mathbb{D}$. Thus, there are lots of pure potentials.

Our properties by themselves do not distinguish functions which are equal almost everywhere. In order to get some of the deeper results of potential theory, we refine this equivalence relation as follows. Let us say that a *positive measure m is of finite energy* if it determines a pure potential, that is, if

$$\left|\int \varphi \, dm\right| \leq \|\varphi\| \|m\|, \quad \varphi \in K(W) \cap \mathbb{D}. \tag{8.5.15}$$

Here $\|m\|$ is simply the norm of the pure potential m determines. We write $\xi_+ =$ set of all positive measures of finite energy.

Let $f \in \mathbb{D}$. Then f is in fact an equivalence class of functions. We select a representative (in fact a class of representatives) v with the following properties:

$v \in L^1(dm)$ for every $m \in \xi_+$; whenever $\varphi_n \in K(W) \cap \mathbb{D}$ converges in $\mathbb{D}$ to f, φ_n converges to v in $L^1(dm)$ for every $m \in \xi_+$.

Clearly, any two representatives with the above properties coincide quasi everywhere, i.e., are equal m-almost everywhere for every $m \in \xi_+$. The procedure for such a selection is the following: Let $f \in \mathbb{D}$ and $\varphi_n \in K(W) \cap \mathbb{D}$ such that $\|\varphi_n - f\|$ tends to zero and

$$\sum \|\varphi_n - \varphi_{n+1}\| \leq \infty. \tag{8.5.16}$$

Since $\|(|\varphi_n - \varphi_{n+1}|)\| \leq \|\varphi_n - \varphi_{n+1}\|$, from (8.5.15) and (8.5.16),

$$\sum \int |\varphi_n - \varphi_{n+}| \, dm < \infty, \quad m \in \xi_+. \tag{8.5.17}$$

In particular (because $|(\Re x)^+ - (\Re y)^+| \leq |x-y|$, etc.), $(\Re \varphi_n)^+$, $(\Re \varphi_n)^-$, etc. all converge in $L^1(dm)$ and m-almost everywhere for all $m \in \xi_+$. Define a representative v of f by

$$v = \liminf (\Re \varphi_n)^+ - \liminf (\Re \varphi_n)^- + i \text{ (similarly)}.$$

Clearly, $v \in L^1(dm)$ for every $m \in \xi_+$. If ψ_n is a sequence in $\mathbb{D} \cap K(W)$ converging in $\mathbb{D}$ to f then by (8.5.15), ψ_n is a Cauchy sequence in $L^1(dm)$. Also $\|\varphi_n - \psi_n\| \to 0$, which by (8.5.15) and what we have said above means that ψ_n tends to v in $L^1(dm)$.

From now on, we assume that we have made such a selection. In particular, we may assume that every $f \in \mathbb{D}$ is in $L^1(dm)$ for every $m \in \xi_+$. A simple consequence of all this is as follows: Let $m \in \xi_+$ determine the pure potential Gm (just notation) then

$$(Gm, f) = \int \overline{f} \, dm, \quad f \in \mathbb{D}, \tag{8.5.18}$$

and in particular, $\|Gm\|^2 = \int Gm \, dm$ (recall $Gm = |Gm|$).

Now we can prove the main principles of potential theory in our setup. We give two illustrations.

Corollary 8.5.6. *The domination principle is valid in $\mathbb{D}$: Let m and n be positive measures of finite energy with potentials u and v. If $u \leq v$, m-almost everywhere then the inequality holds everywhere.*

Proof. By Corollary 8.5.5, $w = u \wedge v$ is the potential of a measure l. By (8.5.18),

$$\|u - w\|^2 = \int u \, dm - \int w \, dm - \int u \, dl + \int w \, dl \leq 0$$

because $u = w$ m-almost everywhere and $w \leq u$ everywhere. Thus $w = u$. □

Corollary 8.5.7. *The equilibrium principle is valid in $\mathbb{D}$: Let F be a compact set which supports a measure of finite energy. Then there is a measure m with support F whose potential $Gm = 1$ quasi everywhere on F and $Gm \leq 1$ everywhere on W.*

Proof. The set of probability measures on F with finite energy is a closed convex subset of $\mathbb{D}$ as is clear from (8.5.18). Let m be the probability measure with minimal energy with corresponding pure potential u.

The restriction of a measure $n \in \xi_+$ to any set A is itself in ξ_+: For every $v \in \mathbb{D}$, $|\int v\, dn| \leq \int |v|\, dn \leq \|(|v|)\|\, \|n\| \leq \|v\|\, \|n\|$. This and an argument very similar to that of Proposition 8.1.3 show that $u = \|u\|^2$ m-almost everywhere and $u \geq \|u\|^2$ quasi everywhere on F. $\|u\|^{-2} m$ gives rise to a pure potential v which is equal to 1 m-almost everywhere and is larger than or equal to 1 quasi-everywhere on F. By the domination principle, it is sufficient to show that $v \wedge 1$ is a pure potential.

$Bz = \inf\{1, (\Re z)^+\}$ being a normal contraction, $v \wedge 1 \in \mathbb{D}$. The set $X = \{h \in \mathbb{D} : \Re h \geq v \wedge 1\}$ is a closed convex set with a unique element f of minimal norm. Corollary 8.5.5 easily implies that f is a pure potential. $f \wedge 1$ belongs to X and being a contraction of f its norm cannot exceed that of f. By minimality, $f = f \wedge 1$. □

The following simple lemma is found useful.

Lemma 8.5.8. *In a Dirichlet space $\mathbb{D}$, the set of pure potentials is a closed convex set (in particular it is complete). And the linear span of pure potentials is dense in $\mathbb{D}$.*

Proof. The first part is trivial from Proposition 8.5.4. For the second, note that by Property 2 of the definition of a Dirichlet space, every bounded non-negative measurable function h with compact support determines a pure potential Gh by the prescription $(f, Gh) = \int fh$. Therefore, the only element in $\mathbb{D}$ orthogonal to all pure potentials is zero. □

Some knowledge of Fourier transforms will be assumed in the following example. An excellent reference (also for the terminology used here) is Chapter 7 of Rudin [48]. Thus, $D =$ set of C^∞-functions with compact supports.

Example. Let σ be a positive measure on $\mathbb{R}^d$ such that

$$\int \left(|x|^2 \wedge 1\right) \sigma\,(dx) < \infty \qquad (8.5.19)$$

and let ψ be defined by

$$\psi(\alpha) = 2\int [1 - \cos(\alpha, x)]\sigma(dx), \quad \alpha \in \mathbb{R}^d. \tag{8.5.20}$$

Provide D with the norm

$$\|f\|^2 = \int \sigma(dy) \int |f(x+y) - f(x)|^2 \, dx. \tag{8.5.21}$$

it is clear how to define an inner product (f, g) so that $\|f\|^2 = (f, f)$. Denote by $\mathbb{D}$ the set of functions f such that f is the almost everywhere finite limit of a sequence of functions in D, the said sequence being at the same time a Cauchy sequence in the norm given by (8.5.21). *With the understanding that two functions are considered equal if they are equal almost everywhere, $\mathbb{D}$ is a Dirichlet space provided ψ^{-1} is locally integrable, where ψ is defined in* (8.5.20).

Let us first show that (8.5.21) defines a finite quality. This is easy to see directly but we use Fourier transforms because we need this later. Applying Parseval to the inner integral in (8.5.21) we can write

$$\|f\|^2 = \text{const} \cdot \int \sigma(dy) \int |\hat{f}(\alpha)|^2 |1 - \exp(i(\alpha, x))|^2 \, d\alpha \tag{8.5.22}$$

$$= \text{const} \cdot \int |\hat{f}|^2(\alpha) \psi(\alpha) \, d\alpha$$

where the constant is $(2\pi)^{-d}$. Now ψ is easily seen to be continuous. The simple inequality $1 - \cos(\alpha, x) \le |\alpha|^2 |x|^2 \wedge 1 \le |\alpha|^2 (|x|^2 \wedge 1)$ shows, using (8.5.19), that $\psi(\alpha) = O(|\alpha|^2)$ at ∞. Since the Fourier transforms in (8.5.22) decrease rapidly at ∞, (f, g) is well defined.

More generally for $f \in \mathbb{D}$, (8.5.21) makes sense: Suppose $\varphi_n \in \mathbb{D}$ converge almost everywhere to f and is a Cauchy sequence in the norm given by (8.5.21). Then for every y, $\lim(\varphi_n(x+y) - \varphi_n(x)) = f(x+y) - f(x)$ almost all x or by Fubini, this limit relation holds for $(dxd\sigma)$-almost all pairs (x, y). Thus because φ_n is a Cauchy sequence, $\|f\| < \infty$ and $\|f - \varphi_n\|$ tends to zero.

With these simple preliminaries it is easy to show that normal contractions operate on $\mathbb{D}$: Let $f \in \mathbb{D}$ and T a normal contraction (see (8.5.11)). There is a Cauchy sequence $\varphi_n \in \mathbb{D}$ such that φ_n tends to f almost everywhere and $\{\varphi_n(x+y) - \varphi_n(x)\}$ tends $dxd\sigma$-almost everywhere to $f(x+y) - f(x)$. Since T is continuous $T\varphi_n$ tends to Tf almost everywhere.

Choose non-negative $a_n \in D$ with $\int a_n = 1$ such that a subsequence of $b_n = (T\varphi_n) * a_n$ tends to Tf almost everywhere.

[This can be dome as follows. Let m be a finite measure equivalent to Lebesgue measure on $\mathbb{R}^d$. Then $T\varphi_n$ tends to Tf in m-measure. There are a's in D such that $T\varphi_n * a$ is uniformly close to $T\varphi_n$. So for suitable a_n, b_n will tend to Tf in m-measure. A subsequence will then tend m-almost everywhere to Tf and m is equivalent to the Lebesgue measure.] b_n belongs to D and by Fubini $b_n(x+y) - b_n(x)$ tends $dxd\sigma$-almost everywhere to $Tf(x+y) - Tf(x)$. Also

$$\|b_n\| \leq \|T\varphi_n\| \quad \text{and} \quad |T\varphi_n(x+y) - T\varphi_n(x)| \leq |\varphi_n(x+y) - \varphi_n(x)|.$$

The first inequality is a simple consequence of Schwarz inequality (note that $0 \leq a_n$ and $\int a_n = 1$) and Fubini, the second is just (8.5.11). Since φ_n is a Cauchy sequence, the second inequality above implies that $|T\varphi_n(x+y) - T\varphi_n(x)|^2$ is uniformly integrable (namely the integral over sets of small measure in uniformly small) and so by Exercise 4.1.7, $\|T\varphi_n - Tf\|$ tends to zero. But then Fatou Lemma, together with this last fact and the first of the above inequalities says that $\|b_n\|$ tends to $\|Tf\|$. And again by the same Exercise, $\|b_n - Tf\|$ tends to zero. In particular b_n is a Cauchy sequence. So $Tf \in \mathbb{D}$. Of course by (8.5.11) and (8.5.21), $\|Tf\| \leq \|f\|$.

To show that $\mathbb{D}$ is a Dirichlet space, we need only prove that the defining Property 2 is valid and that $\mathbb{D}$ is complete. The latter follows from the former because it would imply that a Cauchy sequence in $\mathbb{D}$ converges locally in $L^1(\mathbb{R}^d)$. Let us prove Property 2. Let $A \in D$ be such that its Fourier transform $\hat{A}$ is strictly positive [If $b \in D$ is real and symmetric, i.e., $b(x) = b(-x)$ and $a = b * b$, then $\hat{a} \geq 0$. If p is the standard Gauss Kernel, $p * a$ is strictly positive and its Fourier transform $= \hat{p} \cdot \hat{a} \in D$.] Then for any $f \in D$, if g is the Fourier transform of $|f|$,

$$\int |f| \hat{A} = \int gA \leq \left(\int |g|^2 \psi\right)^{1/2} \left(\int \psi^{-1} |A|^2\right)^{1/2}. \tag{8.5.23}$$

The last member of (8.5.23) is finite because A has compact support and ψ^{-1} is locally integrable. And the third member is simply the norm in $\mathbb{D}$ of $|f|$ — because (8.5.22) is valid for any $f \in L^2(\mathbb{R}^d)$ — and from what we have already shown ($z \mapsto |z|$ is a normal contraction) this norm does not exceed the norm of f. Since A is strictly positive, Property 2 follows from (8.5.23). Thus $\mathbb{D}$ *is a Dirichlet space*.

By (8.5.22), the map $f \mapsto \hat{f}$ is an isometry (up to a constant which in our context is unimportant) of D into $L^2(\psi)$. By the density of D in $\mathbb{D}$, this extends to an isometry of $\mathbb{D}$ into $L^2(\psi)$. To show that it is actually onto, it is sufficient to show that the image of $\mathbb{D}$ is dense: namely that the set of $\hat{f}$ with $f \in D$ is dense in $L^2(\psi)$. If $g \in L^2(\psi)$ and $\int \bar{g}\hat{f}\psi = 0$ for all $f \in D$, then if A is as in (8.5.23), $\int \overline{\hat{f}}\hat{g}\hat{f}\psi = 0$ for all $f \in D$. Since $\bar{g}\hat{A}\psi$ is in L^1, this is equivalent by Parseval to $\int \widehat{\bar{g}\hat{A}\psi} f = 0$ for all $f \in D$, i.e., that $\bar{g}\hat{A}\psi = 0$ almost everywhere. But $\hat{A} > 0$ and from (8.5.20), ψ can only have isolated zeroes and so g must be zero. *Thus the map $f \mapsto \hat{f}$ of $\mathbb{D}$ into $L^2(\psi)$ extends to an isometry — up to a constant — of $\mathbb{D}$ onto $L^2(\psi)$.* This fact leads to the following useful description of $\mathbb{D}$:

Theorem 8.5.9. *A locally summable function f belongs to $\mathbb{D}$ if and only if there is a $g \in L^2(\psi)$ such that for all $\varphi \in D$ we have $\int f\varphi = (2\pi)^{-d} \int g\bar{\hat{\varphi}}$. The $\mathbb{D}$-norm of f is $(2\pi)^{-d/2}$ times the $L^2(\psi)$-norm of g. We are justified in calling g the Fourier transform of f.*

Proof. Note that the theorem implicitly claims that $g\hat{\varphi}$ is summable for every $g \in L^2(\psi)$ and $\varphi \in D$. Let T be the map inverse to the map $\varphi \mapsto \hat{\varphi}$ of D into $L^2(\psi)$. T is continuous on $L^2(\psi)$ onto $\mathbb{D}$. Let $a \in D$. The map $f \mapsto \int fa$ is continuous on $\mathbb{D}$ by Property 2. So there is an element $b \in L^2(\psi)$ such that $\int Tg\, a = \int g\bar{b}\psi$ for all $g \in L^2(\psi)$. Now if $\varphi \in D$ and $g = \hat{\varphi}$ then $Tg = \varphi$ so that $\int \varphi a = \int \hat{\varphi}\bar{b}\psi$ for all $\varphi \in D$. By Parseval, this is the same as $\int \hat{\varphi}\bar{\hat{a}} = (2\pi)^d \int \hat{\varphi}\bar{b}\psi$ for all $\varphi \in D$. As before, $\hat{a} = (2\pi)^d b\psi$. Since T is 1-1 onto, one part of Theorem 8.5.9 follows.

What we have proved above can be restated as follows: For every $a \in D$, and $g \in L^2(\psi)$, $g\hat{a}$ is summable (indeed $\hat{a} = b\psi$ for some $b \in L^2(\psi)$ as we saw above), the map $g \mapsto \int g\hat{a}$ is continuous and $\int Tg\, a = (2\pi)^{-d} \int g\hat{a}$. So if f satisfies the conditions of the theorem f must equal Tg. $\square$

Remark. That $\hat{a} = b\psi$ for some $b \in L^2(\psi)$, proved above says in particular that $\int |\hat{a}|^2 \psi^{-1} < \infty$ for each $a \in D$. This implies a certain growth restriction on ψ^{-1}, i.e., that it is a tempered distribution.

A special case. Riesz potentials. We describe the situation briefly and invite the reader to supply the details. For simplicity, we ignore constants, i.e., equations hold up to multiplicative constants.

Let $0 < \theta < 2$ and $\sigma(dy) = |y|^{-d-\theta}$. The function $\psi(\alpha)$ (8.5.20) is a constant multiple of $|\alpha|^\theta$. Let $\mathbb{D}$ be the corresponding Dirichlet space.

Consider the Riesz kernels $|x|^{-d+\theta}$ introduced in the beginning of Chapter 4. Using the representation of these in terms of the Brownian semi-group, it is seen that the Fourier transform (in the sense of distributions) of $|x|^{-d+\theta}$ is $|x|^{-\theta}$.

The θ-potential $Im = I(\theta, m)$ of a positive measure m is $Im = m * |x|^{-d+\theta}$; its θ-energy norm is $(\int Im \, dm)^{1/2}$. A complex Radon measure m will be said to have finite energy if the corresponding total variation measure $|m|$ has finite energy. Using the convolution relation between the Riesz kernels established at the beginning of Chapter 4, it is seen that the energy norm of m is the L^2-norm of $I(\frac{\theta}{2}, m)$. Since the Fourier transform (in the distribution sense) of the θ-potential of a finite complex measure is $\hat{m}|\alpha|^{-\theta}$ we see that a finite complex measure m has finite energy if and only if $I(\frac{\theta}{2}, m) \in L^2$, i.e., if and only if $\int |\hat{m}|^2 |\alpha|^{-\theta} < \infty$, which is identical to $\widehat{Im} \in L^2(\psi)$ where $\psi(\alpha) = |\alpha|^\theta$. By Theorem 8.5.9, $Im \in \mathbb{D}$. And the energy norm of m is the L^2-norm of $I(\frac{\theta}{2}, m)$ which by Parseval is $\int |\hat{m}|^2 |\alpha|^{-\theta}$, i.e., the energy norm of m is the $\mathbb{D}$-norm of Im. It is a simple step now to remove the restriction of finiteness of m and we can say: A complex Radon measure m has finite energy if and only if its potential Im belongs to $\mathbb{D}$.

Now let us describe the pure potentials in $\mathbb{D}$. Suppose a finite positive measure m generates a pure potential f. This means $(f, A) = \int \overline{A} \, dm$ for all $A \in \mathbb{D}$ with $(\cdot, \cdot)$ denoting inner product in $\mathbb{D}$; see (8.5.18). For any $\varphi \in D$ we have $\int Im \overline{\varphi} = \int I\overline{\varphi} \, dm$ and, $I\varphi \in \mathbb{D}$ so that $\int Im \overline{\varphi} = (f, I\varphi)$. It is clear from (8.5.22) that scalar product in $\mathbb{D}$ corresponds to scalar product in $L^2(\psi)$. Transferring to $L^2(\psi)$-space this equality reads: if g corresponds to f then $\int Im\varphi = (f, I\varphi) = \int g\widehat{I\varphi}\psi = \int g\overline{\hat{\varphi}}$ because $\psi = |\alpha|^\theta$ and $\widehat{I\varphi} = |\alpha|^{-\theta} \hat{\varphi}$. By Theorem 8.5.9 the last integral is just $\int f\varphi$. Thus, $\int Im\varphi = \int f\varphi$ for all $\varphi \in D$, i.e., $f = Im$. We have proved the following:

$f \in \mathbb{D}$ is a pure potential if and only if $f = Im$ for a positive measure m of finite energy.

By Lemma 8.5.8, we have the following corollary:

Corollary 8.5.10. *The set of positive measures of finite θ-energy is complete under the energy norm. The minimum of two θ-potentials is a θ-potential.*

For different proofs of the first part of Corollary 8.5.10, see [25, pp. 80–94] and [44, pp. 82–90]. For a different proof of the second part, see Landkof [44, p. 129].

Exercises

8.5.1. Show that $f, g \in$ BLD and $f = g$ almost everywhere implies $f = g$ quasi everywhere.

Hint: The set of measures m having density is dense in the set of all measures of finite energy. So if $\int |f - g|\, dm = 0$ for every m with density the same holds for all m of finite energy.

8.5.2. Show that a continuous function equal almost everywhere to a BLD function is itself BLD.

Hint: Follow the proof of step 3 of Lemma 8.5.1.

8.5.3. Let f be BLD and $\epsilon > 0$. Show that there is an open set U of capacity smaller than ϵ such that $f|F$ is continuous where $F = W \backslash U$.

Hint: Let a_n be C^∞ with compact supports such that (the energy norm) $\|a_n - f\| < n^{-2}$. Recall the following: The capacity of any Borel set is the supremum of the capacities of its compact sets; the capacity of a compact set = the total mass of its equilibrium distribution = the energy of its equilibrium distribution. Let $A_n = \{|a_n - f| > n^{-1}\}$. For any compact subset F of A_n and $m =$ the equilibrium distribution of F,

$$n^{-1} m(F) \leq \int |a_n - f|\, dm \leq \|m\| \|a_n - f\| \leq n^{-2} \sqrt{m(F)}$$

showing that $N(A_n)$ ($=$ capacity of A_n) $\leq n^{-2}$. Off $B_k = \bigcup_{n \geq k} A_n$, a_n converges uniformly to f, $N(B_k)$ is small for large k and the capacity of a Borel set if the infimum of the capacities of open sets containing it.

8.5.4. Show that the gradient of the potential of a positive measure m exists almost everywhere and is locally summable. The energy norm of a positive measure is, up to a constant, the L^2-norm of the gradient of its potential.

Hint: That the gradient in the ordinary sense exists almost everywhere and is locally summable follows from Exercise 6.3.1. Energy norm of m is the L^2-norm of the generalized gradient of its potential.

8.5.5. Show that a BLD function f is absolutely continuous on almost all lines parallel to the coordinate axes.

8.5.6. Let $\mathbb{D}$ be as in the Example. Show that if f is integrable, has compact support and $\|f\|$ as defined in (8.5.21) is finite, then $f \in \mathbb{D}$.

Hint: If $0 \leq a_n \in D$ has small support around 0, $\int a_n = 1$, then $f * a_n \in D$, tends to f almost everywhere and $\|f * a_n\| \leq \|f\|$. Now use Fatou and Exercise 4.1.7.

8.5.7. If $\mathbb{D}$ is as in the Example, $f \in \mathbb{D}$ and $\varphi \in D$, then $\varphi * f \in \mathbb{D}$.
Hint: If $\|\varphi_n - f\|$ tends to zero so does $\|\varphi_n * f - \varphi * f\|$.

For Martin boundary, fine topology, and more on balayage, consult [12].

References for Further Study

These are references in addition to the ones already given: [2, 4–7, 11, 13, 15, 16, 18–22, 27, 34, 36–40, 42, 43, 53].

References

[1] K. Adhikari and N. K. Reddy. Hole probabilities for finite and infinite Ginibre ensembles. *Int. Math. Res. Not. IMRN*, (21): 6694–6730, 2017.

[2] N. Agram, S. Haadem, B. Øksendal and F. Proske. Optimal stopping, randomized stopping, and singular control with general information flow. *Theory Probab. Appl.*, 66(4): 601–612, 2022. Translation of *Teor. Veroyatn. Primen.*, 6(6): 760–773, 2021.

[3] Y. Ameur, C. Charlier, J. Cronvall and J. Lenells. Exponential moments for disk counting statistics at the hard edge of random normal matrices. *J. Spectr. Theory*, 13(3): 841–902, 2023.

[4] A. Ancona, R. Lyons and Y. Peres. Crossing estimates and convergence of Dirichlet functions along random walk and diffusion paths. *Ann. Probab.*, 27(2): 970–989, 1999.

[5] O. E. Barndorff-Nielsen, F. E. Benth, J. Pedersen and A. E. D. Veraart. On stochastic integration for volatility modulated Lévy-driven Volterra processes. *Stochastic Process. Appl.*, 124(1): 812–847, 2014.

[6] O. E. Barndorff-Nielsen and A. Shiryaev. *Change of Time and Change of Measure*, 2nd edn. Advanced Series on Statistical Science & Applied Probability, Vol. 21. World Scientific Publishing Co. Pte. Ltd., Hackensack, 2015.

[7] H. Bauer. *Harmonische Räume und ihre Potentialtheorie*. Lecture Notes in Mathematics, No. 22. Springer-Verlag, Berlin, 1966. Ausarbeitung einer im Sommersemester 1965 an der Universität Hamburg gehaltenen Vorlesung.

[8] M. L. Bedini and M. Hinz. Credit default prediction and parabolic potential theory. *Stat. Probab. Lett.*, 124: 121–125, 2017.

[9] T. Bloom, N. Levenberg and F. Wielonsky. Logarithmic potential theory and large deviation. *Comput. Methods Funct. Theory*, 15(4): 555–594, 2015.

[10] T. Bloom, N. Levenberg, V. Totik and F. Wielonsky. Modified logarithmic potential theory and applications. *Int. Math. Res. Not. IMRN*, (4): 1116–1154, 2017.

[11] R. M. Blumenthal and R. K. Getoor. *Markov Processes and Potential Theory*. Pure and Applied Mathematics, Vol. 29. Academic Press, New York, 1968.

[12] M. Brelot. *On Topologies and Boundaries in Potential Theory.* Lecture Notes in Mathematics, Vol. 175. Springer-Verlag, Berlin, 1971. Enlarged edition of a course of lectures delivered in 1966.

[13] D. I. Cartwright, P. M. Soardi and W. Woess. Martin and end compactifications for non-locally finite graphs. *Trans. Am. Math. Soc.*, 338(2): 679–693, 1993.

[14] C. Charlier. Hole probabilities and balayage of measures for planar coulomb gases, 2023.

[15] S. Y. Cheng, L.-F. Tam and T. Y.-H. Wan. Harmonic maps with finite total energy. *Proc. Am. Math. Soc.*, 124(1): 275–284, 1996.

[16] C. Constantinescu and A. Cornea. *Potential Theory on Harmonic Spaces.* Die Grundlehren der mathematischen Wissenschaften, Band 158. Springer-Verlag, New York, 1972. With a preface by H. Bauer.

[17] J. B. Conway. *Functions of One Complex Variable. II.* Graduate Texts in Mathematics, Vol. 159. Springer-Verlag, New York, 1995.

[18] J. Deny. Méthodes hilbertiennes en théorie du potentiel. In *Potential Theory (C.I.M.E., I Ciclo, Stresa, 1969)*, Centro Internazionale Matematico Estivo (C.I.M.E.), pp. 121–201. Cremonese, Rome, 1970.

[19] J. L. Doob. Classical potential theory and Brownian motion. In *Aspects of Contemporary Complex Analysis. Proceedings of NATO Advanced Study Institute, University of Durham, Durham, 1979*, pp. 147–179. Academic Press, London, 1980.

[20] J. L. Doob. Kolmogorov's early work on convergence theory and foundations. *Ann. Probab.*, 17(3): 815–821, 1989.

[21] J. L. Doob. *Stochastic Processes.* Wiley Classics Library. John Wiley & Sons, Inc., New York, 1990. Reprint of the 1953 original, A Wiley-Interscience Publication.

[22] J. L. Doob. *Classical Potential Theory and Its Probabilistic Counterpart.* Classics in Mathematics. Springer-Verlag, Berlin, 2001. Reprint of the 1984 edition.

[23] P. D. Dragnev, B. Fuglede, D. P. Hardin, E. B. Saff and N. Zorii. Condensers with touching plates and constrained minimum Riesz and Green energy problems. *Constr. Approx.*, 50(3): 369–401, 2019.

[24] P. D. Dragnev, B. Fuglede, D. P. Hardin, E. B. Saff and N. Zorii. Constrained minimum Riesz energy problems for a condenser with intersecting plates. *J. Anal. Math.*, 140(1): 117–159, 2020.

[25] N. du Plessis. *An Introduction to Potential Theory.* University Mathematical Monographs, No. 7. Hafner Publishing Co., Darien, CT; Oliver and Boyd, Edinburgh, 1970.

[26] J. Eells and B. Fuglede. *Harmonic Maps between Riemannian Polyhedra.* Cambridge Tracts in Mathematics, Vol. 142. Cambridge University Press, Cambridge, 2001. With a preface by M. Gromov.

[27] M. El Kadiri, A. Aslimani and S. Haddad. On the integral representation of the nonnegative superharmonic functions in a balayage space. *Riv. Mat. Univ. Parma (N.S.)*, 10(1): 1–24, 2019.

[28] M. El Kadiri and B. Fuglede. The Dirichlet problem at the Martin boundary of a fine domain. *J. Math. Anal. Appl.*, 457(1): 179–199, 2018.

[29] B. Fuglede. Symmetric function kernels and sweeping of measures. *Anal. Math.*, 42(3): 225–259, 2016.

[30] B. Fuglede. Fine potential theory. In *Potential Theory — Surveys and Problems (Prague, 1987)*. Lecture Notes in Mathematics, Vol. 1344, pp. 81–97. Springer, Berlin, 1988.

[31] B. Fuglede. Capacity, convexity, and isoperimetric inequalities. *Expos. Math.*, 11(5): 455–464, 1993.

[32] B. Fuglede. The Dirichlet problem for harmonic maps from Riemannian polyhedra to spaces of upper bounded curvature. *Trans. Am. Math. Soc.*, 357(2): 757–792, 2005.

[33] B. Fuglede and N. Zorii. Various concepts of Riesz energy of measures and application to condensers with touching plates. *Potential Anal.*, 53(4): 1191–1223, 2020.

[34] L. L. Helms. *Introduction to Potential Theory*. Pure and Applied Mathematics, Vol. XXII. Wiley-Interscience [A division of John Wiley & Sons, Inc.], New York, 1969.

[35] L. L. Helms. *Potential Theory*, 2nd edn. Universitext. Springer, London, 2014.

[36] J. Hoffmann-Jørgensen. *The Theory of Analytic Spaces*. Various Publications Series, No. 10. Aarhus Universitet, Matematisk Institut, Aarhus, 1970.

[37] Y. Hu and B. Øksendal. Linear Volterra backward stochastic integral equations. *Stochastic Process. Appl.*, 129(2): 626–633, 2019.

[38] K. Itô and H. P. McKean, Jr. *Diffusion Processes and Their Sample Paths*. Die Grundlehren der mathematischen Wissenschaften, Band 125. Academic Press, Inc., Publishers, New York; Springer-Verlag, Berlin, 1965.

[39] B. N. Khabibullin. Poisson-Jensen formulas and balayage of measures. *Eurasian Math. J.*, 12(4): 53–73, 2021.

[40] B. N. Khabibullin and E. B. Menshikova. Balayage of measures with respect to polynomials and logarithmic kernels on the complex plane. *Lobachevskii J. Math.*, 42(12): 2823–2833, 2021.

[41] P. Kim, R. Song and Z. Vondraček. On potential theory of Markov processes with jump kernels decaying at the boundary. *Potential Anal.*, 58(3): 465–528, 2023.

[42] P. Kim, R. Song and Z. Vondraček. Potential theory of Dirichlet forms degenerate at the boundary: The case of no killing potential. *Math. Ann.*, 388(1): 511–542, 2024.

[43] K. Kuratowski. *Topology*, Vol. I. Academic Press, New York; Państwowe Wydawnictwo Naukowe [Polish Scientific Publishers], Warsaw, 1966. New edition, revised and augmented, Translated from the French by J. Jaworowski.

[44] N. S. Landkof. *Foundations of Modern Potential Theory*. Die Grundlehren der mathematischen Wissenschaften, Band 180. Springer-Verlag, New York, 1972. Translated from the Russian by A. P. Doohovskoy.

[45] C. Miao, R. Shen and T. Zhao. Scattering theory for the subcritical wave equation with inverse square potential. *Selecta Math. (N.S.)*, 29(3): 30, 2023. Paper No. 44.

[46] M. Mikelashvili. Potential method in the coupled linear quasi-static theory of thermoelasticity for double porosity materials. *Arch. Mech. (Arch. Mech. Stos.)*, 75(5): 559–590, 2023.

[47] W. Rudin. *Real and Complex Analysis*. McGraw-Hill Book Co., New York, 1966.

[48] W. Rudin. *Functional Analysis*. McGraw-Hill Series in Higher Mathematics. McGraw-Hill Book Co., New York, 1973.

[49] S.-M. Seo. Edge behavior of two-dimensional Coulomb gases near a hard wall. *Ann. Henri Poincaré*, 23(6): 2247–2275, 2022.

[50] G. Serafin. On potential theory of hyperbolic Brownian motion with drift. *Probab. Math. Statist.*, 40(1): 1–22, 2020.

[51] B. Skinner. *Logarithmic Potential Theory on Riemann Surfaces*. ProQuest LLC, Ann Arbor, 2015. Thesis (Ph.D.)–California Institute of Technology.

[52] N. T. Varopoulos. *Potential Theory and Geometry on Lie Groups*. New Mathematical Monographs, Vol. 38. Cambridge University Press, Cambridge, 2021.

[53] N. Zorii. On the theory of capacities on locally compact spaces and its interaction with the theory of balayage. *Potential Anal.*, 59(3): 1345–1379, 2023.

Appendix A
Kernels and More General Classes of Gaussian Processes

The present treatment of both stochastic calculus and potential theory, and related themes has its focus on the case when the Gaussian process is the standard Brownian, and its realization as a Wiener process. The purpose of this appendix is to offer an extended framework. We extend the context in two ways. The first way has its starting point the most general positive definite kernel, K realized on the product $X \times X$, where X is a set. We then show that there is then a canonical family of centered Gaussian processes W which have K as covariance kernel. The second framework is focused on the case when K has the form $\mu(A \cap B)$ for A and B in a subset of a prescribed sigma-algebra. This setting allows a realization of an associated white noise calculus. Our aim is to present an extension of Ito calculus to this more general framework [1–3, 7–10, 13, 15, 16, 18].

In this appendix, we present two results. We first note that a centered Gaussian process W may be constructed for an arbitrary positive definite kernel K such that W has K as its covariance kernel. We then present the special case when K is defined from a positive sigma-finite measure μ, i.e., $K(A, B) := \mu(A \cap B)$.

In the latter case, the corresponding Gaussian process is a generalized Brownian motion. Its quadratic variation is μ. In particular, the standard Brownian motion arises for the special case when μ is Lebesgue measure on $\mathbb{R}$.

A.1 Positive Definite Kernels

The concept of a positive definite (p.d.) kernel has emerged as a fundamental tool in various fields of both pure and applied mathematics. This concept extends the ideas of positive definite functions and matrices, offering a more general framework. The development of p.d. kernels can be traced back to the early 20th century, finding applications in diverse areas such as integral operator equations (J. Mercer), harmonic analysis and complex domain theory (G. Szegő and S. Bergmann), and boundary value problems for partial differential equations (PDEs) (N. Aronszajn). Notably, Aronszajn's introduction of the reproducing kernel Hilbert space (RKHS) concept has been pivotal in this field.

In recent times, the application of p.d. kernels has continued to expand, influencing numerous mathematical disciplines. This includes, but is not limited to, the areas initially mentioned. A significant new application area is RKHS theory, which has seen substantial development.

Positive definite kernels and their reproducing kernel Hilbert spaces

Consider a set X and a complex-valued function K on $X \times X$. We say that K is *positive definite* (p.d.) if, for all finite subset F of X and and any set of complex numbers $(\xi_x)_{x \in F}$, the following inequality holds:

$$\sum_{x \in F} \sum_{y \in F} \bar{\xi}_x \xi_y K(x, y) \geq 0 \tag{A.1.1}$$

In other words, the $|F| \times |F|$ matrix $(K(x,y))_{F \times F}$ is positive definite in the usual sense of linear algebra.

The Aronszajn theory of reproducing kernel Hilbert spaces (RKHS), denoted as, denoted $\mathscr{H}(K)$, is essential for our discussion:

For a fixed finite set F and vectors $(\xi_x)_{x \in F}$, $(\eta_x)_{x \in F}$ in $\mathbb{C}^{|F|}$, consider the set of all functions on X of the form

$$\sum_{x \in F} \xi_x K(\cdot, x). \tag{A.1.2}$$

The inner product and norm of such functions are defined as follows:

$$\left\langle \sum_{x \in F} \xi_x K(\cdot, x), \sum_{y \in F} \eta_y K(\cdot, y) \right\rangle_{\mathscr{H}(K)} := \sum_{F \times F} \sum \bar{\xi}_x \eta_y K(x, y), \tag{A.1.3}$$

$$\left\| \sum_{x \in F} \xi_x K(\cdot, x) \right\|^2_{\mathscr{H}(K)} := \sum_{F \times F} \sum \bar{\xi}_x \xi_y K(x, y). \tag{A.1.4}$$

Then $\mathscr{H}(K)$ is the Hilbert completion of all functions as in (A.1.2) with respect to the $\|\cdot\|_{\mathscr{H}(K)}$-norm.

Appendix A: Kernels and More General Classes of Gaussian Processes   235

With the definition of the RKHS $\mathscr{H}(K)$, we get directly that the functions $\{K(\cdot, x)\}_{x \in X}$ are automatically in $\mathscr{H}(K)$, and that, for all $h \in \mathscr{H}(K)$, we have

$$\langle K(\cdot, x), h \rangle_{\mathscr{H}(K)} = h(x), \tag{A.1.5}$$

i.e., the *reproducing* property holds.

Further note that, given K, the RKHS $\mathscr{H}(K)$ is determined uniquely, up to isometric isomorphism in Hilbert space.

Remark A.1.1. In general, when a positive definite kernel K is given, and the reproducing kernel Hilbert space (RKHS) $\mathscr{H}(K)$ is constructed, one only specifies a generating (dense span) system of elements in $\mathscr{H}(K)$. As outlined, this subspace (in $\mathscr{H}(K)$) may be specified as in (A.1.2), and the $\|\cdot\|_{\mathscr{H}(K)}$-norm is then specified as in (A.1.4). With this point of view, $\mathscr{H}(K)$ arises as a Hilbert-completion. However, in many applications it is possible to give an explicit formula for *all* the elements ψ in $\mathscr{H}(K)$, and their Hilbert norms $\|\psi\|_{\mathscr{H}(K)}$.

Lemma A.1.2. *Let $X \times X \xrightarrow{K} \mathbb{C}$ be a p.d. kernel, and let $\mathscr{H}(K)$ be the corresponding RKHS (see (A.1.3)–(A.1.5)). Let h be a function defined on X; then TFAE:*

1. $h \in \mathscr{H}(K)$;
2. *there is a constant $C = C_h < \infty$ such that, for all finite subset $F \subset X$, and all $(\xi_x)_{x \in F}$, $\xi_x \in \mathbb{C}$, the following a priori estimate holds:*

$$\left| \sum_{x \in F} \xi_x h(x) \right|^2 \leq C_h \sum_{x \in F} \sum_{y \in F} \overline{\xi}_x \xi_y K(x, y). \tag{A.1.6}$$

Proof. The implication (1)⇒(2) is immediate, and in this case, we may take $C_h = \|h\|^2_{\mathscr{H}(K)}$.

Now for the converse, assume (2) holds for some finite constant. On the $\mathscr{H}(K)$-dense span in (A.1.2), define a linear functional

$$L_h \left(\sum_{x \in F} \xi_x K(\cdot, x) \right) := \sum_{x \in F} \xi_x h(x). \tag{A.1.7}$$

From the assumption (A.1.6) in (2), we conclude that L_h (in (A.1.7)) is a well defined bounded linear functional on $\mathscr{H}(K)$. Initially, L_h is only defined on the span (A.1.2), but by (A.1.6), it is bounded, and so extends uniquely by $\mathscr{H}(K)$-norm limits. We may therefore apply Riesz' lemma to the Hilbert space $\mathscr{H}(K)$, and conclude that there is a unique $H \in \mathscr{H}(K)$ such that

$$L_h(\psi) = \langle \psi, H \rangle_{\mathscr{H}(K)} \tag{A.1.8}$$

for all $\psi \in \mathscr{H}(K)$. Now, setting $\psi(\cdot) := K(\cdot, x)$, for $x \in X$, we conclude from (A.1.8) that $h(x) = H(x)$; and so $h \in \mathscr{H}(K)$, proving (1). $\square$

A.2 Gaussian Processes

The study of positive definite (p.d.) functions is rooted in three main areas: (i) Fourier and harmonic analysis and (ii) optimization and approximation problems, including spline approximations as developed by I. Schönberg; and (iii) stochastic processes.

We will focus on the third area, stochastic processes. A stochastic process is a collection of random variables indexed by a set, often a group G, such as $\mathbb{R}$ for real-time processes or $\mathbb{Z}$ for discrete-time processes. A key aspect of analyzing stochastic processes is understanding their associated covariance functions.

A process $\{X_g \mid g \in G\}$ is called *Gaussian* if each random variable X_g follows a Gaussian distribution. For Gaussian processes, knowing the first two moments (mean and variance) is sufficient. If we normalize, setting the mean equal to 0, the process is fully characterized by its covariance function. In general, the covariance function is a function on $G \times G$, or on a subset, but if the process is *stationary*, the covariance function will in fact be a p.d. function defined on G, or a subset of G. For a systematic study of positive definite functions on groups G, on subsets of groups, and the variety of the extensions to p.d. functions on G.

By a theorem of Kolmogorov [11], every Hilbert space may be realized as a (Gaussian) reproducing kernel Hilbert space (RKHS), see Theorem A.2.1, and also [6, 14, 17].

Now every positive definite kernel is also the covariance kernel of a Gaussian process; a fact which is a point of departure in our present analysis: Given a positive definite kernel, we explore its use in the analysis of the associated Gaussian process; and vice versa.

This point of view is especially fruitful when one is dealing with problems from stochastic analysis. Even restricting to stochastic analysis, we have the exciting area of applications to statistical learning theory.

Let $(\Omega, \mathscr{F}, \mathbb{P})$ be a *probability space*, i.e., Ω is a fixed set (sample space), $\mathscr{F}$ is a specified sigma-algebra (events) of subsets in Ω, and $\mathbb{P}$ is a probability measure on $\mathscr{F}$.

Appendix A: Kernels and More General Classes of Gaussian Processes

A Gaussian random variable is a function $V : \Omega \to \mathbb{R}$ (in the real case), or $V : \Omega \to \mathbb{C}$, such that V is measurable with respect to the sigma-algebra $\mathscr{F}$ on Ω, and the corresponding sigma-algebra of Borel subsets in $\mathbb{R}$ (or in $\mathbb{C}$). Let $\mathbb{E}$ denote the expectation defined from $\mathbb{P}$, i.e.,

$$\mathbb{E}(\cdots) = \int_\Omega (\cdots)\, d\mathbb{P}. \tag{A.2.1}$$

The requirement on V is that its distribution is Gaussian. If g denotes a Gaussian on $\mathbb{R}$ (or on $\mathbb{C}$), the requirement is that

$$\mathbb{E}(f \circ V) = \int_{\mathbb{R}(\text{or } \mathbb{C})} f\, dg; \tag{A.2.2}$$

or equivalently

$$\mathbb{P}(V \in B) = \int_B dg = g(B) \tag{A.2.3}$$

for all Borel sets B.

If $N \in \mathbb{N}$, and $V_1, \ldots, V_N$ are random variables, the Gaussian requirement is that the joint distribution of $(V_1, \ldots, V_N)$ is an N-dimensional Gaussian, say g_N, so if $B \subset \mathbb{R}^N$ then

$$\mathbb{P}((V_1, \ldots, V_N) \in B) = g_N(B). \tag{A.2.4}$$

For our present purpose we may restrict to the case where the mean (of the respective Gaussians) is assumed zero. In that case, a finite joint distribution is determined by its covariance matrix. In the $\mathbb{R}^N$ case, it is specified as follows (the extension to $\mathbb{C}^N$ is immediate) $(G_N(j_1, j_2))_{j_1, j_2 = 1}^N$,

$$G_N(j_1, j_2) = \int_{\mathbb{R}^N} x_{j_1} x_{j_2} g_N(x_1, \ldots, x_N)\, dx_1 \cdots dx_N \tag{A.2.5}$$

where $dx_1 \cdots dx_N = \lambda_N$ denotes the standard Lebesgue measure on $\mathbb{R}^N$.

The following is known:

Theorem A.2.1 (Kolmogorov [12], see also [4, 5]). *A kernel $K : X \times X \to \mathbb{C}$ is positive definite if and only if there is a (mean zero) Gaussian process $(V_x)_{x \in X}$ indexed by X such that*

$$\mathbb{E}\left(\overline{V}_x V_y\right) = K(x, y) \tag{A.2.6}$$

where $\overline{V}_x$ denotes complex conjugation.

Moreover, the process in (A.2.6) *is uniquely determined by the kernel* K *in question. If* $F \subset X$ *is finite, then the covariance kernel for* $(V_x)_{x \in F}$ *is* K_F *given by*

$$K_F(x,y) = G_F(x,y), \qquad (A.2.7)$$

for all $x, y \in F$, *see* (A.2.5).

In the subsequent sections, we address a number of properties of Gaussian processes important for their stochastic calculus. Our analysis deals with both the general case, and particular examples from applications. We begin in Section A.3 with certain Wiener processes which are indexed by sigma-finite measures. For this class, the corresponding p.d. kernel has a special form; see (A.3.1) in Definition A.3.1.

A.3 Sigma-Finite Measure Spaces and Gaussian Processes

We consider functions of σ-finite measure space $(M, \mathscr{F}_M, \mu)$ where M is a set, $\mathscr{F}_M$ a σ-algebra of subsets in M, and μ is a positive measure defined on $\mathscr{F}_M$. It is further assumed that there is a countably indexed $(A_i)_{i \in \mathbb{N}}$ s.t. $0 < \mu(A_i) < \infty$, $M = \cup_i A_i$, and further that the measure space $(M, \mathscr{F}_M, \mu)$ is complete; so the Radon–Nikodym theorem holds. We also restrict to the case when μ is assumed non-atomic.

Definition A.3.1. Set

$$\mathscr{F}_{\text{fin}} = \{A \in \mathscr{F}_M \mid 0 < \mu(A) < \infty\}.$$

Note then

$$K^{(\mu)}(A,B) = \mu(A \cap B), \ A, B \in \mathscr{F}_{\text{fin}} \qquad (A.3.1)$$

is positive definite. The corresponding Gaussian process $(W_A^{(\mu)})_{A \in \mathscr{F}_{\text{fin}}}$ is called the Wiener process. In particular, we have

$$\mathbb{E}\left(W_A^{(\mu)} W_B^{(\mu)}\right) = \mu(A \cap B) \qquad (A.3.2)$$

and

$$\lim_{(A_i)} \sum_i \left(W_{A_i}^{(\mu)}\right)^2 = \mu(A). \qquad (A.3.3)$$

The precise limit in (A.3.3), quadratic variation, is as follows: Given μ as above, and $A \in \mathscr{F}_{\text{fin}}$, we then take limit over the filter of all partitions of

Appendix A: Kernels and More General Classes of Gaussian Processes 239

A (see (A.3.4)) relative to the standard notation of refinement:

$$A = \cup_i A_i, \quad A_i \cap A_j = \emptyset \text{ if } i \neq j, \text{ and } \lim \mu(A_i) = 0. \tag{A.3.4}$$

Details: Let $(\Omega, Cyl, \mathbb{P})$, $\mathbb{P} = \mathbb{P}^{(\mu)}$ be the probability space which realizes $W^{(\mu)}$ as a Gaussian process (or generalized Wiener process), i.e., s.t. (A.3.2) holds for all pairs in $\mathscr{F}_{\text{fin}}$. In particular, we have that $W_A^{(\mu)} \underset{\text{(dist)}}{\sim} N(0, \mu(A))$, i.e., mean zero, Gaussian, and variance $= \mu(A)$. Then, we have the following:

Lemma A.3.2. *With the assumptions as above, we have*

$$\lim_{(A_i)} \mathbb{E}\left(\left|\mu(A)\mathbb{1} - \sum_i (W_{A_i}^{(\mu)})^2\right|^2\right) = 0 \tag{A.3.5}$$

where (in (A.3.5)) the limit is taken over the filter of all partitions (A_i) of A, and $\mathbb{1}$ denotes the constant function "one" on Ω.

As a result, we get the following Ito integral

$$W^{(\mu)}(f) := \int_M f(s)\, dW_s^{(\mu)}, \tag{A.3.6}$$

defined for all $f \in L^2(M, \mathscr{F}, \mu)$, and

$$\mathbb{E}\left(\left|\int_M f(s)\, dW_s^{(\mu)}\right|^2\right) = \int_M |f(s)|^2\, d\mu(s). \tag{A.3.7}$$

We note that the following operator,

$$L^2(M, \mu) \ni f \longmapsto W^{(\mu)}(f) \in L^2(\Omega, \mathbb{P}) \tag{A.3.8}$$

is isometric.

In our subsequent considerations, we shall need the following precise formula (see Lemma A.3.3) for the RKHS associated with the p.d. kernel

$$K^{(\mu)}(A, B) := \mu(A \cap B), \tag{A.3.9}$$

defined on $\mathscr{F}_{\text{fin}} \times \mathscr{F}_{\text{fin}}$. We denote the RKHS by $\mathscr{H}(K^{(\mu)})$.

Lemma A.3.3. *Let μ be as above, and let $K^{(\mu)}$ be the p.d. kernel on $\mathscr{F}_{\text{fin}}$ defined in (A.3.9). Then the corresponding RKHS $\mathscr{H}(K^{(\mu)})$ is as follows: A function Φ on $\mathscr{F}_{\text{fin}}$ is in $\mathscr{H}(K^{(\mu)})$ if and only if there is a $\varphi \in L^2(M, \mathscr{F}_M, \mu) \,(=: L^2(\mu))$ such that*

$$\Phi(A) = \int_A \varphi\, d\mu, \tag{A.3.10}$$

for all $A \in \mathscr{F}_{\text{fin}}$. Then

$$\|\Phi\|_{\mathscr{H}(K^{(\mu)})} = \|\varphi\|_{L^2(\mu)}. \tag{A.3.11}$$

Proof. To show that Φ in (A.3.10) is in $\mathscr{H}(K^{(\mu)})$, we must choose a finite constant C_Φ such that, for all finite subset $(A_i)_{i=1}^N$, $A_i \in \mathscr{F}_{\text{fin}}$, $\{\xi_i\}_{i=1}^N$, $\xi_i \in \mathbb{R}$, we get the following *a priori* estimate:

$$\left| \sum_{i=1}^N \xi_i \Phi(A_i) \right|^2 \leq C_\Phi \sum_i \sum_j \xi_i \xi_j K^{(\mu)}(A_i, A_j). \tag{A.3.12}$$

But a direct application of Schwarz to $L^2(\mu)$ shows that (A.3.12) holds, and for a finite C_Φ, we may take $C_\Phi = \|\varphi\|_{L^2(\mu)}^2$, where φ is the $L^2(\mu)$-function in (A.3.10). The desired conclusion now follows from an application of Lemma A.1.2.

We have proved one implication from the statement of the lemma: Functions Φ on $\mathscr{F}_{\text{fin}}$ of the formula (A.3.10) are in the RKHS $\mathscr{H}(K^{(\mu)})$, and the norm $\|\cdot\|_{\mathscr{H}(K^{(\mu)})}$ is as stated in (A.3.11). In the below, we shall denote these elements in $\mathscr{H}(K^{(\mu)})$ as pairs (Φ, φ). We shall also restrict attention to the case of real valued functions.

For the converse implication, let H be a function on $\mathscr{F}_{\text{fin}}$, and assume $H \in \mathscr{H}(K^{(\mu)})$. Then by Schwarz applied to $\langle \cdot, \cdot \rangle_{\mathscr{H}(K^{(\mu)})}$ we get

$$\left| \langle H, \Phi \rangle_{\mathscr{H}(K^{(\mu)})} \right| \leq \|H\|_{\mathscr{H}(K^{(\mu)})} \|\varphi\|_{L^2(\mu)}, \tag{A.3.13}$$

where we used (A.3.11). Hence when Schwarz is applied to $L^2(\mu)$, we get a unique $h \in L^2(\mu)$ such that

$$\langle H, \Phi \rangle_{\mathscr{H}(K^{(\mu)})} = \int_M h\, \varphi\, d\mu \tag{A.3.14}$$

for all (Φ, φ) as in (A.3.10). Now specialize to $\varphi = \chi_A$, $A \in \mathscr{F}_{\text{fin}}$, in (A.3.14) and we conclude that

$$H(A) = \int_A h\, d\mu; \tag{A.3.15}$$

which translates into the assertion that the pair (H, h) has the desired form (A.3.10). And hence by (A.3.11) we have $\|H\|_{\mathscr{H}(K^{(\mu)})} = \|h\|_{L^2(\mu)}$ as stated. This concludes the proof of the converse inclusion. $\square$

References

[1] N. Attia and A. Akgül. On solutions of biological models using reproducing Kernel Hilbert space method. In *Computational Methods for Biological Models*. Studies in Computational Intelligence, Vol. 1109, pp. 117–136. Springer, Singapore, 2023.

[2] J. Baek and J.-S. Chen. A neural network-based enrichment of reproducing kernel approximation for modeling brittle fracture. *Comput. Methods Appl. Mech. Eng.*, 419: 24, 2024. Paper No. 116590.

[3] M. Engliš and E.-H. Youssfi. M-harmonic reproducing kernels on the ball. *J. Funct. Anal.*, 286(1): 54, 2024. Paper No. 110187.

[4] T. Hida. *Brownian Motion*. Applications of Mathematics, Vol. 11. Springer-Verlag, New York, 1980. Translated from the Japanese by the author and T. P. Speed.

[5] T. Hida. Stochastic variational calculus. In *Stochastic Partial Differential Equations and Their Applications (Charlotte, NC, 1991)*. Lecture Notes in Control and Information Sciences, Vol. 176, pp. 123–134. Springer, Berlin, 1992.

[6] K. Itô and H. P. McKean, Jr. *Diffusion Processes and Their Sample Paths*. Die Grundlehren der Mathematischen Wissenschaften, Band 125. Academic Press Inc., Publishers, New York, 1965.

[7] P. Jorgensen and J. Tian. Harmonic analysis of network systems via kernels and their boundary realizations. *Discrete Contin. Dyn. Syst. Ser. S*, 16(2): 277–308, 2023.

[8] P. E. T. Jorgensen, M.-S. Song and J. Tian. Infinite-dimensional stochastic transforms and reproducing kernel Hilbert space. *Sampl. Theory Signal Process. Data Anal.*, 21(1): 27, 2023. Paper No. 12.

[9] T. Karvonen. Small sample spaces for Gaussian processes. *Bernoulli*, 29(2): 875–900, 2023.

[10] M. Katori and T. Shirai. Zeros of the i.i.d. Gaussian Laurent series on an annulus: Weighted Szegő kernels and permanental-determinantal point processes. *Commun. Math. Phys.*, 392(3): 1099–1151, 2022.

[11] A. N. Kolmogorov. On logical foundations of probability theory. In *Probability Theory and Mathematical Statistics (Tbilisi, 1982)*. Lecture Notes in Mathematics, Vol. 1021, pp. 1–5. Springer, Berlin, 1983.

[12] A. N. Kolmogorov and J. A. Rozanov. On a strong mixing condition for stationary Gaussian processes. *Teor. Verojatnost. i Primenen.*, 5: 222–227, 1960.

[13] I. M. Park, S. Seth, M. Rao and J. C. Príncipe. Strictly positive-definite spike train kernels for point-process divergences. *Neural Comput.*, 24(8): 2223–2250, 2012.

[14] K. R. Parthasarathy and K. Schmidt. Stable positive definite functions. *Trans. Am. Math. Soc.*, 203: 161–174, 1975.

[15] L. G. Sanchez Giraldo, M. Rao and J. C. Principe. Measures of entropy from data using infinitely divisible kernels. *IEEE Trans. Inform. Theory*, 61(1): 535–548, 2015.

[16] L. Santilli and M. Tierz. Riemannian Gaussian distributions, random matrix ensembles and diffusion kernels. *Nucl. Phys. B*, 973: 30, 2021. Paper No. 115582.

[17] B. S. Nagy, C. Foias, H. Bercovici and L. Kérchy. *Harmonic Analysis of Operators on Hilbert Space*. Universitext. Springer, New York, 2010. Enlarged edition.

[18] J. Wang and M. Hillman. Upwind reproducing kernel collocation method for convection-dominated problems. *Comput. Methods Appl. Mech. Eng.*, 420: 2024. Paper no. 116711.

Index

B

balayage, 129, 184, 189, 197–198, 209–210, 213

Brownian motion, 9, 15, 19–20, 23, 25–26, 34, 43, 47, 54, 84, 233

C

capacity, 184, 198–199, 201, 205–206, 209, 217, 228

conditionally independent, 13

D

Dirichlet problem, 71, 78, 81, 86, 137, 149

Dirichlet space, 184, 214, 220

Doob's upcrossing inequality, 7–8

Dynkin's formula, 16, 71, 178

E

Evans-Vascilesco, 133

G

Gaussian processes, 233, 236

Green function, 60, 143–144, 151, 153, 156, 169, 171, 173, 178, 186

L

Laplacian, 15, 53, 173

M

Markov property, 19–20, 24, 33, 72, 114, 190

martingale, 5, 7, 11, 15, 19, 114, 140

P

positive definite kernel, 233, 238

potential theory, 183, 186, 233

R

Riesz measure, 123, 125–129, 135, 154, 190, 199, 205, 214, 216

right continuous, 13–14, 139

S

sampling, 5, 7, 114

semi groups, 43–47, 49, 51, 53–54, 57, 61, 67

spherical harmonics, 72, 98, 100, 103

stopping time, 6–7, 13, 15, 26, 72, 112, 114

sub-martingale, 6, 11, 14

subharmonic, 115, 118, 120, 128, 165

super martingale, 6–7, 10, 13, 114

superharmonic functions, 109–113, 116, 118, 121, 123–125, 127, 130, 133, 145, 147, 154, 156, 158, 167, 180, 216

U

uniform integrability, 10, 12

V

vague convergence, 196, 217